The Rightful Place of Science:
Biofuels

The Rightful Place of Science: Biofuels

John A. Alic

forward by

Clark A. Miller

Consortium for Science, Policy and Outcomes
Tempe, AZ and Washington, DC

THE RIGHTFUL PLACE OF SCIENCE: BIOFUELS

For information, Rightful Place of Science series

Consortium for Science, Policy and Outcomes

PO Box 875603, Tempe, AZ 85287-5603

Model citation for this volume:

Alic, John A. 2013. The Rightful Place of Science: Biofuels. Tempe, AZ: Consortium for Science, Policy and Outcomes.

The Rightful Place of Science series explores the complex interactions among science, technology, politics and the human condition.

Other volumes in this series:

Zachary, G.P., ed. 2013. The Rightful Place of Science: Politics. Tempe, AZ: Consortium for Science, Policy and Outcomes.

ISBN: 0615903428

ISBN-13: 978-0615903422

LCCN: 2013920545

First Edition, November 2013

ACKNOWLEDGMENTS

This work was completed in partnership with the Consortium for Science, Policy and Outcomes and the Clean Air Task Force, supported through a grant by the Nathan Cummings Foundation.

CONTENTS

FIGURE

TABLES

ACRONYMS

AFCO	Alternative Fuel Certification Office (US Air Force)
ARPA-E	Advanced Research Projects Agency-Energy
Btu	British thermal unit
CAFE	Corporate Average Fuel Economy
CO_2	carbon dioxide
DARPA	Defense Advanced Research Projects Agency
DLA Energy	Defense Logistics Agency Energy
DoD	Department of Defense
DOE	Department of Energy
DPA	Defense Production Act
EIA	Energy Information Administration
EISA	Energy Independence and Security Act
EPA	Environmental Protection Agency
FT	Fischer-Tropsch
GHG	greenhouse gas
GMR	giant magnetoresistance
GTL	gas-to-liquid
HDD	hard disk drive
HEFA	hydroprocessed esters and fatty acids
HRJ	hydroprocessed renewable jet
JP	jet propellant
LNG	liquefied natural gas
MOD	Ministry of Defence (United Kingdom)
MTG	methanol-to-gasoline
RFS	Renewable Fuel Standard
SPK	synthetic paraffinic kerosene
tfc	trillion cubic feet
USDA	US Department of Agriculture

FORWARD

Clark A. Miller *

The next 50 years will bring a fundamental transformation of the world's energy systems. Already that change is beginning. Drilling for unconventional oil and gas is exploding, in the Marcellus Shale, in the tar sands of Alberta, and in the deepest waters of the Gulf of Mexico. New investments in solar and wind energy have made renewable energy the fastest growing sector of the energy industry. China and India are leading a new nuclear renaissance. A gold rush is on for the energy technologies that will power the global economy by mid-century. And the stakes are vast. Today's global energy market is already $5 trillion. Tomorrow's will be even bigger.

For those who worry about climate change—which seems to be almost everyone these days—a central question for energy policy is how to address the future of fuels. Carbon-based fuels have huge advantages over other fuels, especially in terms of their energy density and their ubiquity. With new fracking technologies, oil and gas are seemingly everywhere and easy and cheap, albeit environmentally destructive, to get out of the ground. Unfortunately, carbon-based fuels are also deeply unsustainable, driving massive changes in the

* Clark A. Miller is Associate Director of the Consortium for Science, Policy and Outcomes at Arizona State University.

Earth's climate and putting much of the planet at risk of severe climatic disruptions.

Averting a climate disaster will ultimately require a solution for fuels drawn from very limited options: hydrogen, carbon capture, or the creation of renewable fuels. Of these, the only option based on currently available technology is the last: biofuels. Indeed, biofuels have important advantages over other options for replacing more traditional fuels. Biofuels would require relatively minor changes to automobile technologies and the infrastructures used to fuel them, especially in comparison to other alternatives, such as hydrogen and carbon capture. Indeed, much of the gasoline purchased in the United States already contains partial concentrations of biofuel, in the form of ethanol produced from corn.

Not surprisingly, therefore, policy leaders have targeted biofuels as a key renewable energy option in the United States and other places with large agricultural sectors, like Brazil and the European Union. Biofuels offer a potentially promising avenue for raising agricultural revenues and thus strengthening rural economies. Over the past several decades, countries have pursued a variety of policies to advance both the production of biofuels and the development of new biofuels technologies and markets. Those policies include research investments, fuel composition requirements, and advanced biofuels directives. Yet, these policies have largely failed to deliver the goods. Worse, they have contributed to problematic side effects, including rising global prices for food.

Biofuels policy thus stands at a stark crossroads. Should biofuels play an important role in efforts to create a sustainable energy future? Will the biofuels sector be able to achieve the ambitious policy goals set for it? How will large-scale biofuels production impact both agro-ecosystems and agricultural markets?

John Alic's book *Biofuels* is essential reading for those interested in these questions confronting the biofuels industry, its policy allies, and its opponents. This short book reviews the history of US policy efforts to support the development of biofuels, the successes and failures of those efforts, and the complexities of launching a US biofuels industry. The book's most important lesson lies in the last of these three. Creating a new energy industry is one of the most complex enterprises human societies have ever sought to undertake. Perhaps a hundred years ago, the challenge was less—then, again, perhaps not, if the stories told about Standard Oil, Schlumberger, and others are even half true.

Today, biofuels innovators face the combined vagaries of five major systems that cut across social, technological, economic, environmental, and policy domains. One is the nation's innovation policy system, in which multiple, diverse federal agencies invest in research and development, each according to their own often ad hoc and highly idiosyncratic reasons. The second is the nation's innovation financing system, comprised of diverse sources of capital—federal agencies, large companies, banks, venture capitalists, angel investors—that innovators must navigate at different stages in the birth and growth of new firms. The third is the energy system itself, which demands extremely high volume/low cost production of commodity fuels to be competitive with existing fuels and the companies that produce them. This forces start-up companies to navigate myriad non-fuel markets as they seek strategies to generate profit while scaling up their business. The fourth is the agricultural system, and its ancillary systems, like water. US agriculture is responsible for the nation's (and a not insignificant amount of the world's) food production and intersects prospective biofuels markets at numerous points, including landowners, land, fertilizers, transportation, feedstocks, processing facilities, labor, and many more. The final is the larger planetary ecosystem, which must accommodate a second major domain of industrialized biological production while already under the strains of accommodating the massive demands of global agricultural production. Crafting a thriving new biofuels industry demands finding integrated solutions that work simultaneously across all of these systems.

This book shines in highlighting how these diverse systems come together to create the context in which a nascent biofuels industry finds itself struggling to survive amidst multiple pressures coming to bear from very different directions. It is hardly surprising, therefore, that US policies have fallen short of their ambitious goals. Are biofuels the answer to sustainable fuels of the future? I don't know, nor does anyone else at this point, whatever supposed truths they may be peddling. What is clear is that, unless the US energy sector comes to grip with the full range of complexities of creating a sustainable energy future—as John Alic describes here—better US biofuels policies are unlikely to either occur or be of much help.

1

INTRODUCTION

From the time a century ago that refined petroleum products first found large markets, policymakers have worried that oil supplies might run out, crippling their nation's militaries along with economic sectors that depend on liquid fuels.

> *We are ... confronted with [oil] price movements of a far-reaching character, which are, I apprehend, part of an attempt on a gigantic scale to corner the market and to control the output.*

As a result,

> *Naval planners everywhere seemed to be gripped with a severe case of nerves while trying to figure out how they were going to procure adequate fuel supplies at reasonable price.*[1]

As the rhetorical flourishes indicate, these are not the

[1] John A. DeNovo, "Petroleum and the United States Navy before World War I," *Mississippi Valley Historical Review*, Vol. 41, No. 4 (March 1955), pp. 641-656; quotation on p. 649.

words of a contemporary government official. They were spoken by Winston Churchill, and not at the time of World War II but in 1913 when he was First Lord of the Admiralty. New discoveries and higher oil recovery rates forestalled the emergency Churchill feared.

Almost any type of liquid fuel—gasoline, diesel fuel, jet fuel—can be made from coal, natural gas, or biomass; all that is needed is a feedstock rich in carbon and hydrogen. From the 1920s, when early processes for making synthetic fuels were being developed, costs have posed the chief obstacle, not in absolute terms but relative to petroleum. Much synthetic fuels technology, then and now, is more or less standard in the chemical industry. Sometimes indeed it might seem as if chemists and chemical engineers can make almost any final product starting with almost any feedstock. Such is not the case; yet for hydrocarbons it is not far from the truth. For synthetic fuels, the question has always been how to compete with petroleum. That is as true of biofuels, liquids produced from biomass—corn or post-harvest residue ordinarily left in fields, dedicated energy crops, urban wastes otherwise destined for incineration or a landfill—as it is of synfuels made from coal, which got most of the attention until a few years ago.

Today it is not simply dollar costs that matter, but environmental costs, especially those associated with global warming. Unfortunately, environmental costs have not yet begun to be priced in dollars. They range beyond greenhouse gases (GHGs) such as carbon dioxide (CO_2) released when the fuel is burned. It takes energy to produce fuel, and more energy to produce synthetics than to refine crude oil into petroleum fuels. These energy inputs must be considered too, including the energy costs and environmental consequences of cultivating, collecting, transporting, and processing biomass.

Many companies, most of them small and entrepreneurial, have been working to commercialize and scale up production of biofuels, especially cellulosic ethanol, made from the cell walls of plants and thus not a food source. Under the schedule Congress set in the 2007 Energy Independence and Security Act (EISA), cellulosic ethanol is to supplement corn ethanol, the principal biofuel consumed in the United States today. Corn ethanol, which does compete with food crops, contributes little or nothing to GHG reduction and barely meets the sustainability test of positive energy returns, that is, energy content in the fuel supplied to final customers exceeding that consumed in production of the fuel. Technical advances in processing cellulosic ethanol, which promises somewhat larger energy returns, have been slow, fuel has yet to reach the market, and the legislated EISA schedule has been put off.

Reserves of petroleum and natural gas, meanwhile, hardly seem exhausted. US production has increased over the past few years, with imports declining, a consequence of technological innovations including advances in geophysical exploration methods and horizontal drilling and hydraulic fracturing, or fracking. As so often in the past, just as synthetic fuels seem poised for innovation and growth, new supplies of fossil fuels appear, now including the prospect of drop-in substitutes for petroleum made from low-cost natural gas. While gas-to-liquid (GTL) fuels are not sustainable, they do promise a near-term alternative to petroleum and a way to begin the transition away from oil. The principal obstacle is the energy consumed in processing, which, depending on conversion efficiencies and the sources of process heat and electrical power consumed, could result in somewhat higher or somewhat lower life-cycle climate impacts.

The US Department of Defense (DoD), the only government agency that counts as a major consumer of fossil fuels, has explored alternatives to petroleum for decades, citing concerns similar to those raised by Churchill before World War I. Because of DoD's deep and wide-ranging R&D and its demanding mission requirements—and because DoD has spurred innovation since World War II in technologies ranging from jet propulsion to digital systems—it gets much attention in the chapters that follow.[2]

Until recently, most DoD work on fuels focused on synthetics made from coal. Although coal is cheap and widely available, greenhouse impacts now rule it out as a source of synfuels and the Pentagon has signaled its intent to begin purchasing biofuels. Jet fuel makes up the bulk of DoD's consumption and the Navy and Air Force have set ambitious near-term goals for purchase of "drop-in" synthetic jet fuel, fully compatible with existing systems and infrastructure. Drop-in alternatives are essential because the military's ships, aircraft, and ground vehicles were designed to run on petroleum, remain in service for decades, and cost far too much to replace. DoD's policy leaves selection of feedstock and processing technology to industry, provided that the fuels meet military standards (which for jet fuel are nearly the same as commercial standards), relying on competition among firms to converge on cost-effective technologies. The approach should foster well-adapted alternatives positioned for ongoing advances through

[2] For more on the reasons, see *Innovation Policy for Climate Change* (Washington, DC: Consortium for Science, Policy and Outcomes, September 2009) and *Energy Innovation at the Department of Defense: Assessing the Opportunities* (Washington, DC: Consortium for Science, Policy and Outcomes, March 2012).

continuous innovation of the sort that has propelled the petrochemical industry for more than a century. At the same time, the policy depends on strong market competition—which, as for cellulosic ethanol, may not appear.

DoD seeks drop-in alternative fuels because it cannot modify or replace its large and costly inventory of existing equipment, much of it decades old. The US motor vehicle fleet numbers nearly 250 million cars and trucks, most of them designed to burn gasoline. On average, these vehicles are old too, if not so old as the B-52 bombers that in 2006 served as initial testbeds for Air Force certification of non-petroleum jet fuels. The motor vehicle fleet has been aging and cars remain on the road longer, slowing turnover, so that it takes years for new technology, such as very high-mileage vehicles, to penetrate. Transportation, along with fossil-fuel burning electrical power plants, contributes the bulk of US CO_2 emissions and, since CO_2 is the major GHG, the bulk of the US contribution to global warming. Federal policies seek to reduce motor vehicle CO_2 emissions through increases in fuel efficiency and alternatives to conventional powertrains (e.g., electric vehicles), an approach that will take decades to make much difference. Drop-in synthetics in place of petroleum—changing the fuel rather than the vehicle—promise nearer-term results while complementing other policies already in place, such as higher fuel economy standards.

Together, diesel fuel and jet fuel, both kerosene, account for one-third of US consumption of transportation fuels and proven GTL technologies exist for making drop-in synthetic kerosene from biomass or natural gas. Production of synthetic gasoline is perfectly possible too, though more costly. Almost regardless of type of biomass, and there are many possibilities, constraints imposed by the need to collect and transport feedstock to biorefineries limit prospects for cost-competitiveness with petroleum. GTL plants fed by natural gas avoid that constrain and now operate in several parts of the world; a single facility the size of Shell's GTL complex in Qatar produces as much kerosene as 100 biodiesel plants of typical US size built in response to EISA.

In his 2012 State of the Union address, President Barack Obama looked forward to domestic reserves of natural gas that could last a century. Natural gas is not sustainable in the long run and certainly no solution to global warming. Indeed, regulations may be needed to ensure that GHG emissions from GTL synthetics do not exceed those for petroleum. But natural gas is available now, and with continuing technical development and appropriate regulations it should be possible to reduce GHG emissions below those for petroleum. Over time, biogas could take the place of natural gas, yielding further GHG reductions. The approach taken by the nation's military, which subsequent chapters explore, could serve, in some respects, as a model for transition to non-petroleum fuels in the civilian economy.

2

FUELS

Hydrocarbons—molecules made up solely of carbon (C) and hydrogen (H) atoms—come in thousands of varieties. Some are small, others, including many of those found in coal, very large. The H and C atoms may be arranged as long chains or closed rings or some combination of these. Chemists and chemical engineers have devised a multitude of processes for breaking hydrocarbon molecules apart and stitching them together, usually in the presence of catalysts. Prime examples include hydrocarbons refined from crude oil with molecular weights appropriate for gasoline or jet fuel, as described in Appendix A1. Synthetic liquid hydrocarbons can also be made from any of a large number of potential biomass feedstocks. Many of the processes start with gasification, producing biogas rich in hydrogen and simple carbon compounds such as carbon monoxide. As for coal gas and natural gas, with further processing biogas can be made into a suitable liquid fuel. There are also synthesis pathways that do not begin with gasification.

Oil and petrochemical firms choose the products they make depending on the prices those products command

and the costs of production. These costs include the prices paid for feedstocks, whether oil, natural gas, or biomass. Periodically, rising oil prices and pessimistic estimates of reserves have spurred enthusiasm for feedstocks that substitute for crude oil. Over the past several decades, recognition of the causes and threats of global warming has enhanced the appeal of biomass as a feedstock. When fossil fuels burn, carbon atoms combine with oxygen in the air to form CO_2, a major cause of global warming. Burning biomass or biofuels also releases CO_2, with this difference: growing the biomass removes CO_2 from the atmosphere, opening the prospect of net-zero or net-negative CO_2 output.[1] Even at oil prices of recent years, however, investors have generally viewed biofuels as too risky. Among the reasons: the prices commanded by petroleum from low-cost fields far outstrip costs of production, so that oil firms can make money even at low prices; for biofuels firms, by contrast, falling oil prices could quickly turn a profitable operation into a money-loser; big oil companies have much political power and the financial resources to undercut or otherwise disadvantage smaller rivals if they wish; and the US government, like oil prices, behaves unpredictably and could abandon policies that subsidize or otherwise support production and consumption of biofuels.

[1] Chapter 2, "Bioenergy," in Ottmar Edenhofer, et al., eds., *Renewable Energy Sources and Climate Change Mitigation: Special Report of the Intergovernmental Panel on Climate Change* (Cambridge, UK: Cambridge University Press, 2012), pp. 209-331, provides an extensive technical, environmental, and economic review. For a short introduction, see Ausilio Bauen, et al., *Bioenergy – A Sustainable and Reliable Energy Source: A Review of Status and Prospects* (N.p.: IEA Bioenergy [International Energy Agency], 2009), Chapter 3, pp. 27-39.

Technically, one of the major obstacles to greater consumption of biofuels such as ethanol has been incompatibility with existing systems and equipment—planes, vehicles, pipelines, and wholesale/retail distribution infrastructure. Essentially all such equipment has been designed for gasoline or kerosene. (As explained in Appendix A1, both jet fuel and diesel fuel are considered kerosenes, even though the average molecular weight of diesel is greater.) Americans, unlike Europeans, have avoided diesel-powered personal vehicles. (Europeans pay much higher fuel taxes and choose more efficient diesels in over half of new car purchases.) Even so, as Figure 1 shows, kerosene accounts for about one-third of US consumption of transportation fuels.

Figure 1. US Consumption of Transportation Fuels, 2012

(domestic production plus imports minus exports).

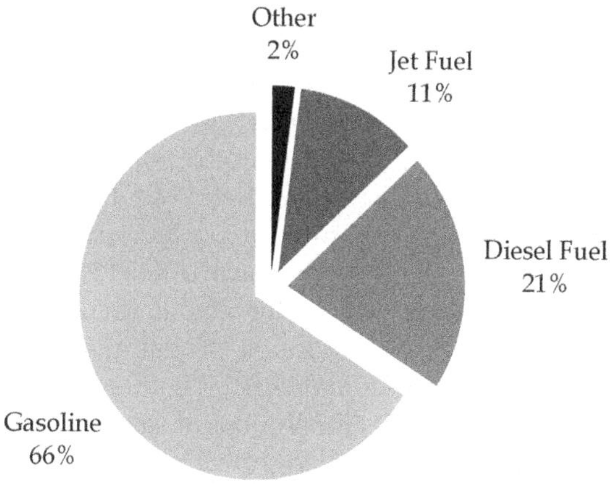

Source: *Monthly Energy Review April 2013*, DOE/EIA-0035(2013/04) (Washington, DC: Energy Information Administration, April 25, 2013), Table 3.7c, p. 55; lubricants excluded.

Fuels and powerplants go together. Years ago automakers began designing engines to run on leadfree gasoline and incorporated minor changes consistent with small percentages of ethanol, which is not a hydrocarbon but an alcohol with the chemical formula C_2H_5OH. Ethanol served as a "fuel extender," added to gasoline in part for its oxygen content, which helps control exhaust emissions, and in part as a replacement for lead compounds to raise octane and suppress efficiency-robbing detonation, or knock, in engines.

The armed forces have large inventories of equipment designed to operate on petroleum, mostly jet fuel, which accounts for nearly 80 percent of DoD purchases by volume (Table 1). Because these inventories would be exorbitantly expensive to replace or even to modify, alternative fuel for the US military means, basically, drop-in synthetic jet fuel—made from feedstocks other than crude oil and meeting the same or similar standards as JP-5 and JP-8, the two main grades purchased by DoD.

Private standards-setting bodies such as ASTM International (formerly the American Society for Testing and Materials) establish and maintain specifications for gasoline, diesel fuel, and jet fuel. Manufacturers of aircraft gas turbines design their powerplants for fuels that meet these specifications and DoD's closely similar standards. JP-5 and JP-8 differ chiefly in a property called flash point, an indicator of flammability. The Navy specifies less flammable JP-5, with a 40 F higher flash point temperature, for shipboard safety. These two fuels, in turn, differ only slightly from the two main commercial grades of jet fuel, Jet A, an ASTM specification widely followed in the United States, and Jet A-1, the standard adopted in most of the rest of the world. JP-8 is in fact the same as commercial Jet A-1 except for extra additives such as corrosion inhibitors and stabilizers (military fuels sometimes sit in storage tanks for lengthy periods). Appendix A2 provides background on fuel specifications.

Table 1

Department of Defense Bulk Fuel Purchases, Fiscal Year 2012 [a]

Fuel Type	*Millions of gallons*
Jet Fuel [b]	3520
JP-8	2002
JP-5	526
Other (Jet A, Jet A-1, and military grades other than JP-8 and JP-5)	992
F-76 and other distillates [c]	1013
Gasoline (aviation and road)	49

[a] Does not include minor accounting categories such as direct purchases at civilian outlets (airports, gas stations). Annual consumption differs from purchases as inventories rise or fall. Some figures rounded.

[b] The main types of jet fuel differ only slightly; see Appendix A2 for fuel types and specifications.

[c] Gas turbines burning F-76 kerosene, heavier and less volatile than jet fuel, power most of the Navy's non-nuclear vessels; diesel engines in some ships likewise burn F-76.

Source: *Defense Logistics Agency Energy Fact Book, Fiscal Year 2012* (Ft. Belvoir, VA: Defense Logistics Agency Energy, n.d.), p. 33.

The Defense Logistics Agency Energy (DLA Energy) buys fuel for all the services, which reimburse the agency via budget transfers. Because the US military maintains hundreds of bases around the world, DLA

Energy purchases fuel on the international market, around half of it abroad, aiming to minimize costs of fuel delivered at the point of consumption.[2] Cargo planes and tankers for air-to-air refueling consume well over half the JP-8 bought for the Air Force, DLA Energy's biggest customer.[3] The services also consume jet fuel for many purposes other than powering aircraft. In 1988 DoD designated JP-8 its battlefield or logistics fuel, expected to be available almost everywhere for almost everything, except at sea. The Army and Marine Corps now burn JP-8 in trucks, armored vehicles, and diesel-powered generators; since much of this equipment had been designed for diesel fuel, considerable testing preceded the shift. Troops in the field also cook and heat water with JP-8. Although purchases and consumption by fuel type vary from year to year depending on inventory fluctuations and operational patterns, JP-8 has been and will continue to be the US military's primary fuel.

Prices for JP-5 and JP-8 track those for civilian grades of jet fuel. Suppliers generally charge slightly more for JP-8 than Jet A-1, with a small additional premium for JP-5.[4] In addition to these high-volume fuels, DLA

[2] Anthony Andrews, *Department of Defense Fuel Spending, Supply, Acquisition, and Policy*, R40459 (Washington, DC: Congressional Research Service, September 22, 2009); Moshe Schwartz, Katherine Blakeley, and Ronald O'Rourke, *Department of Defense Energy Initiatives: Background and Issues for Congress*, R42558 (Washington, DC: Congressional Research Service, December 10, 2012).

[3] *Energy Horizons: United States Air Force Energy S&T Vision, 2011-2026*, AF/ST TR 11-01 (N.p.: US Air Force, January 31, 2012), Fig. 2.2, p. 8. Air Force transport and cargo planes fly personnel and equipment around the world for all the services.

[4] DLA Energy's fiscal 2013 price lists put JP-5 at two cents per

Energy buys many others in smaller quantities. So long as one of the services has a "requirement"—for an exotic liquid rocket propellant or, say, fuel made from algae for R&D purposes—the agency will make the purchase. Spending on bulk fuel in fiscal 2012, the quantities listed in Table 1 plus other items such as lubricants and high-energy missile fuels, totaled about $15 billion.[5] While little changed from 2011, this figure represents a considerable increase over 2009 and 2010, when oil prices were lower, although less than the 2008 peak of $16.5 billion, when oil was high and so was consumption because of intense fighting in both Iraq and Afghanistan.[6] DoD's fuel purchases by volume over the past decade averaged around one-third higher than during the preceding, more peaceful, decade.

The Air Force, responsible within DoD for most fuels R&D, put considerable money into drop-in alternatives made from coal and other domestic fossil resources in the aftermath of the 1973-74 Arab oil embargo. Biofuels are a more recent addition to the agenda. The Air Force transition from JP-4, the predecessor to JP-8, began in

gallon more than JP-8, $3.75 compared to $3.73. "Standard Fuel Prices in Dollars, FY2013 President's Budget, FY2013 Rates," <www.energy.dla.mil/DLA_finance_energy/Pages/dlafp03.aspx>. These are the prices DLA Energy charges the services, which the agency adjusts periodically to keep its books balanced, not the actual purchase prices paid by DoD.

[5] *Defense Logistics Agency Energy Fact Book, Fiscal Year 2012* (Ft. Belvoir, VA: Defense Logistics Agency Energy, n.d.), p. 28. The high-flying U-2, for example, burns a specially formulated kerosene with lower freezing point and enhanced thermal stability at roughly twice the price of JP-8.

[6] *Defense Logistics Agency Energy Fact Book, Fiscal Year 2010* (Ft. Belvoir, VA: Defense Logistics Agency Energy, 2011), p. 25.

1979 and was officially completed only in 1996 (and DoD still buys JP-4 in some volume). Any widespread adoption of synthetics will probably follow a similarly protracted timetable.

3

FIRST-GENERATION AND "ADVANCED" BIOFUELS

Brazil, the United States, and a number of other countries, mostly in Europe, produce synthetic fuels from biomass—corn, sugar cane, soybeans—because of public policies. These include financial subsidies and regulatory mandates that provide guaranteed markets, an indirect subsidy. Were the invisible hand of the market to somehow lock into place oil prices of, say, $200 per barrel, biofuels would be a good business. But that is not how invisible hands work, and biofuels does not look like a good business absent subsidies, except perhaps for risk-taking entrepreneurs and innovators with faith in their technologies.

Oil firms and biofuels firms operate with quite different cost structures. Some oil is remarkably inexpensive to produce. Costs are much higher for other deposits, "tight oil" or reservoirs in the Arctic. At high enough oil prices, biofuels firms expect comfortable profits. But when oil prices fall, petroleum firms with access to low-cost crude still make money; biofuels producers make losses. So long as there are no economic penalties for emitting GHGs, subsidies will be necessary to induce biofuels production unless or until investors believe oil prices will stay high permanently. But the problem is more intractable still. A century of expanding demand for fossil energy created both the scale and learning economies to drive down costs for petroleum. Such conditions are unlikely to pertain for carbon-neutral alternative fuels in the foreseeable future.

To be sure, motives for subsidizing biofuels vary. Countries without domestic petroleum reserves may seek to reduce imports. Politically powerful agricultural interests may lobby for new markets. Global warming is not a first-order concern of DoD, nor was climate change the primary motive for subsidies extended by government in the United States to support production of corn ethanol and biodiesel.

The First Generation: Corn Ethanol and Biodiesel

Estimates by the Energy Information Administration (EIA) put total US government subsidies for biomass and biofuels at $7.8 billion in 2010, most of this for corn ethanol.[1] Many state governments also subsidize corn ethanol and biodiesel.[2] Beginning with the 2005 Energy Policy Act, the federal government required fuel blenders to mix biofuels with gasoline and diesel before sale.[3] By compelling fuel blenders to buy, mandated

[1] *Direct Federal Financial Interventions and Subsidies in Energy in Fiscal Year 2010* (Washington, DC: Energy Information Administration, July 2011), Tables ES2 and 3, pp. xiii and 13.

[2] See, e.g., *Current State of the U.S. Ethanol Industry* (New Castle, DE: Cardno ENTRIX for Department of Energy, November 30, 2010), pp. 9-3 – 9-15.

[3] Randy Schnepf, *Agriculture-Based Biofuels: Overview and*

blending creates an assured market for biofuels. The initial Renewable Fuel Standard (RFS) set the 2006 blending requirement at 4 billion gallons, about 2 percent of road vehicle fuel consumption that year, increasing in subsequent years. Almost immediately Congress accelerated and extended the schedule in the 2007 Energy Independence and Security Act, which established what came to be known as RFS2. Other policies that foster biofuels production and consumption include incentives for automakers to market flexible-fuel vehicles and tax credits for biofuels producers.

Because of the RFS/RFS2 mandate, and because of ethanol's value as a knock-reducing substitute for lead and an oxygenate in controlling exhaust pollution, it has become the primary biofuel produced and sold in the United States. Corn is big business in the Midwest, available in volume for biorefiners to purchase alongside other buyers, and ethanol production is simple, with modest capital requirements. On the other hand, with a net energy gain (the difference between the energy content of the fuel and the energy consumed in cultivation, harvesting, processing, and so on) of no more than 20-30 percent, corn ethanol barely meets the sustainability test of positive energy returns.[4] Although

Emerging Issues, R41282 (Washington, DC: Congressional Research Service, May 1, 2013), provides a concise review.

[4] Jason Hill, Erik Nelson, David Tilman, Stephen Polasky, and Douglas Tiffany, "Environmental, Economic, and Energetic Costs and Benefits of Biodiesel and Ethanol Biofuel," *Proceedings of the National Academy of Sciences,* Vol. 103, No. 30, July 25, 2006, pp. 11206-11210. Biodiesel does better on this measure. Also see, e.g., Charles A.S. Hall, Bruce E. Dale, and David Pimental, "Seeking to Understand the Reasons for Different Energy Return on Investment (EROI) Estimates for Biofuels," *Sustainability,* Vol. 3, 2011, pp. 2413-2432.

many types of powerplants, including jet engines, can be designed to run on alcohol, few of the cars and trucks in the existing US fleet will tolerate more than 10-15 percent ethanol in gasoline. Annual capacity for producing corn ethanol now exceeds 14 billion gallons, more than enough for blending at 10 percent with all the gasoline sold in the United States.[5] Today's "gasohol" contains 10 percent ethanol; newer vehicles will accept 15 percent ethanol and such blends may reach the market in volume within a few years.

RFS2 also covers diesel fuel. Blenders can meet the requirement, set at 1.28 billion gallons for 2013, by purchasing either biodiesel or renewable diesel. These are chemically distinct. "Biodiesel," a somewhat misleading but now accepted label, contains oxygen and for that reason, and like ethanol, is not a hydrocarbon. Renewable diesel, on the other hand, is chemically indistinguishable from petroleum diesel. It is also more expensive to make (and output has been negligible). Soybeans have been the feedstock of choice for biodiesel produced in the United States, which, again like ethanol, reaches final customers blended in small percentages with petroleum diesel. Capacity is only about 15 percent of that for corn ethanol and with many biodiesel plants idle in the past several years, for reasons summarized in Box A, output has been far short of capacity.

Under RFS2, corn ethanol consumption is to reach 15 billion gallons in 2015, after which this figure becomes

[5] Schnepf, *Agriculture-Based Biofuels,* Table 2, p. 15. As of September 2012, the Renewable Fuels Association, counted 211 ethanol plants, most of them operating with corn as feedstock, while the trade association for biodiesel firms listed 166 plants. *Biorefinery Locations,* <www.ethanolrfa.org/bio-refinery-locations/>; *Biodiesel Plants Listing,* <www.biodiesel.org/production/plants/plants-listing>.

the ceiling in order to limit upward pressure on food prices. Until the advent of RFS/RFS2, most of the US corn crop sold for animal feed. Ethanol suppliers took over one-quarter of the harvest in 2006 and well over 40 percent beginning in 2010, pushing up prices for corn with spillover effects on other food products.[6] In 2012, drought in the Midwest cut deeply into corn yields, pushing prices to record highs and at least some ethanol plants into the red. EISA mandates for "advanced" biofuels made from non-food biomass, such as cellulosic ethanol (produced from the cell walls of plant matter), aim to moderate these effects.

Box A

Biodiesel and Renewable Diesel

Under federal regulatory definitions, renewable diesel has the same chemical composition as petroleum diesel; biodiesel, a mixture of three types of long-chain fatty acids known as esters (specifically, fatty acid methyl esters, FAME), does not.[a] FAME molecules contain oxygen and are not hydrocarbons. Because of this, biodiesel properties differ considerably from those of petroleum or renewable diesel. In particular, biodiesel exhibits poor cold flow behavior and poor chemical stability; over time, the constituents may repolymerize, with degradation in properties and possible damage to equipment. For such reasons, biodiesel generally reaches the pump mixed at low percentage with petroleum diesel (e.g., as B5).

The differences between biodiesel and renewable diesel stem from processing, not from feedstocks, which may include plant oils (e.g, from soybeans), animal fats, and other forms of biomass including municipal waste. Biodiesel is

[6] Department of Agriculture, Economic Research Service, <www.ers.usda.gov/media/866543/cornusetable.html>.

made by breaking up triglycerides, the primary constituent of plant oils and animal fats, at low temperatures and pressures, yielding FAME molecules. The equipment is simple (hobbyists make biodiesel at home). Renewable diesel, on the other hand, is made with standard petroleum refining processes, such as hydroprocessing/hydrotreating (see Appendix A1). These processes, capital-intensive, taking place at high temperatures and pressures, and suited to high-volume production, remove the oxygen (combining it with hydrogen to form water vapor) and break down the large and heavy molecules in biomass more-or-less irreversibly. Since the equipment and processes are the same, renewable diesel could in principle be coproduced with petroleum diesel.

The United States produced little biodiesel until 2005, when a $1 per gallon tax credit took effect. Output grew rapidly through 2008, then fell. With biodiesel plants cheap to build and operate, feedstock purchases can account for three-quarters or more of costs, and output responds to both oil prices and feedstock prices. Plants may shut down if crude oil prices, hence diesel prices, fall, or soybean prices rise. Stop-and-start tax credits have also affected output. Congress allowed the $1 per gallon credit to expire at the end of 2009, then renewed it a year later. While the renewal was retroactive, covering all production in 2010, firms that had stopped making biodiesel could not benefit. The tax credit expired again at the end of 2011 and output in 2012 did not reach half of capacity.[b]

[a] *Guidance for Biodiesel Producers and Biodiesel Blenders/Users,* EPA420-B-07-019 (Washington, DC: Environmental Protection Agency, November 2007); *Biodiesel Handling and Use Guide*, Fourth Edition, NREL /TP-540-43672 (N.p.: National Renewable Energy Laboratory, December 2009).

[b] *Monthly Biodiesel Production Report* (Washington, DC: Energy Information Agency, April 2013). Total capacity comes to about 2.1 billion gallons per year.

Cellulosic Ethanol

By creating demand, the Renewable Fuel Standard spurred investment in production capacity for corn ethanol and biodiesel. Even though RFS2 led to "an explosion of new companies, new business relationships, and new advances in the cellulosic biofuel industry" the policy has not as yet worked for advanced biofuels, for lack of production capacity.[7] The Environmental Protection Agency (EPA), with discretion under EISA to alter or waive the RFS2 schedule, has had no choice but to cut the cellulosic ethanol mandate year by year.

Not long after enactment of EISA, the United States, along with much of the rest of the world, entered a deep and persistent recession. Biofuels investments fell precipitously.[8] Even so, the economic slump was only one reason, and not necessarily the major reason, why firms have been reluctant to design and build plants for producing advanced biofuels. Projected costs exceed costs for first-generation biofuels, R&D advances have been disappointing, and future profit streams seem correspondingly elusive, even with the EISA market reserve (Box B). A National Research Council committee, for example, concluded that without subsidies beyond EISA cellulosic ethanol could promise profits only at oil

[7] *Renewable Fuel Standard Program (RFS2) Regulatory Impact Analysis*, EPA-420-R-10-006 (Washington, DC: Environmental Protection Agency, February 2010), p. 173.

[8] Worldwide biofuels investment, more than $26 billion in 2007, totaled perhaps $7 billion in 2011, according to Bloomberg New Energy Finance estimates reported in *ICAO Review: Sustainable Alternative Fuels for Aviation* (Montréal, Canada: International Civil Aviation Organization, n.d. [evidently 2011]), p. 46.

prices in the neighborhood of $200 per barrel, roughly twice 2012 levels.[9] Meanwhile biofuels firms, acutely aware of ongoing attacks on RFS2 by the oil industry and some environmental groups, fear it could disappear.

Box B

Advanced Biofuels and the Renewable Fuel Standard

Under the Energy Independence and Security Act, ethanol made from non-food cellular material, basically the cell walls of plants (e.g., corn stover—post-harvest remnants of the crop ordinarily left in the field), was to become the major US second-generation biofuel, supplementing corn ethanol.[a] EISA set cellulosic ethanol consumption in 2010, the first year of the RFS2 mandate, at 100 million gallons, rising to 3 billion gallons in 2015 and 16 billion gallons in 2022, the last year covered by the law. The market reserve spurred many companies to announce ambitious plans.[b] Few have been fulfilled.

Feedstock appears to be abundant, certainly for the modest quantities scheduled in early years. The proportion of cellular material in agricultural residues and energy crops ranges from perhaps 35 percent to 60 percent. While some post-harvest crop residues should be left in fields to maintain soil quality, often half or more depending on the crop and local conditions, considerable expansion in dedicated energy crops (similar to tree farms supplying the pulp and paper industry) appears feasible. Candidate second-generation feedstocks include fast-growing woody shrubs and grasses. Algae, sometimes termed a third-generation feedstock, does not need even marginally

[9] *Renewable Fuel Standard: Potential Economic and Environmental Effects of U.S. Biofuel Policy* (Washington, DC: National Academies Press, 2011), pp. 121-123. This report lists, pp. 160-161, 28 federal programs directed at reducing production costs for cellulosic ethanol.

productive cropland. The obstacles have been in processing. They are both technical and economic.

Compared to processes for making first-generation corn ethanol (more properly termed corn starch ethanol since the energy-rich sugars come from starch in the kernels), which are simple and inexpensive, methods under consideration for cellulosic ethanol have proved difficult to master. Cellular material must first be separated out so the energy-rich sugars can be extracted. These sugars, released for instance by treatment with enzymes at elevated temperatures, are then fermented and distilled. There are many ways to accomplish these tasks. When EISA took effect, no firm had proven process technology in hand and no one knew which of many candidates might turn out to be best. Many companies invested in R&D. Some built pilot and demonstration plants. Development flagged and investors grew wary. Estimated costs for designing, building, and equipping biorefineries rose: with more steps in processing, front-end expenditures appear to be about three times those for corn ethanol plants.[c] Big companies in the oil and chemical industries have the capital, should they wish to spend it on cellulosic ethanol; financing has been difficult for smaller companies and startups.[d]

Firms small or large have hesitated to scale up from demonstration and pilot facilities. In 2008 DuPont opened a 250,000 gallon-per-year pilot facility in Tennessee; not until late 2012 did the company break ground for a production-scale cellulosic ethanol plant in Iowa.[e] DuPont's is one of the few such announcements and dozens would be needed to meet the RFS2 schedule. It is hard to see how capacity could catch up with the mandate, at least over the next half-dozen years. Both industry and environmental groups have sued the Environmental Protection Agency over its RFS2 decisions and this will probably continue (along with challenges to biofuel blending requirements imposed by state governments). If the opponents of EISA force reconsideration of the RFS2 schedule, investments in plant and equipment for advanced biofuels would be devalued; so would intellectual property stemming from R&D.

Total biofuels consumption under RFS2 is to reach 36 billion gallons in 2022. Most of this would continue to be ethanol, 15 billion gallons from corn and 16 billion gallons from cellulosic biomass. The remaining 5 billion gallons is to be made up of other "advanced" biofuels. Almost any "renewable fuel" other than corn ethanol qualifies, provided it "has lifecycle greenhouse gas emissions ... at least 50 percent less than .... gasoline or diesel (whichever is being replaced by the renewable fuel) ...."[f]

[a] Randy Schnepf and Brent D. Yacobucci, *Renewable Fuel Standard (RFS): Overview and Issues*, R40155 (Washington, DC: Congressional Research Service, March 14, 2013). While RFS2 applies to motor vehicle fuels only, aviation fuels made from biomass earn credits that could be sold to gasoline blenders.

[b] For examples, see, *Renewable Fuel Standard Program (RFS2) Regulatory Impact Analysis*, EPA-420-R-10-006 (Washington, DC: Environmental Protection Agency, February 2010), pp. 113ff, and *Cellulosic Biofuels Industry Progress Report 2012-2013* (N.p.: Advanced Ethanol Council, n.d. [evidently December 2012]).

[c] Kelsi Bracmort, *Meeting the Renewable Fuel Standard (RFS) Mandate for Cellulosic Biofuels: Questions and Answers*, R41106 (Washington, DC: Congressional Research Service, January 11, 2012), p. 4.

[d] Shikar Ghosh and Ramana Nanda, "Venture Capital Investment in the Clean Energy Sector," Harvard Business School Working Paper 11-020, November 2010, pp. 7-12.

[e] Jim Collins, "DuPont: A Leader in Biofuels," presentation at Biotechnology Industry Organization briefing, National Press Club, Washington, DC, September 5, 2012.

[f] *Energy Independence and Security Act of 2007*, Title II, Sec. 201, <www.gpo.gov/fdsys/pkg/BILLS-110hr6enr/.../BILLS-110hr6enr.pdf >.

The original Renewable Fuel Standard created a powerful stimulus for production of corn ethanol, and RFS2 reinforced it. Unfortunately, alcohols such as ethanol cannot substitute for gasoline in existing vehicles. Moreover, when alcohols burn in air to produce energy in the form of heat, the oxygen, effectively dead weight, makes no contribution. As a result, the energy density of ethanol—useful energy released in combustion per gallon or per pound of fuel—is less than that of hydrocarbons (Table 2). Butanol, a heavier alcohol with less oxygen relative to carbon and hydrogen (C_4H_9OH) has better properties, but butanol, unlike ethanol and methanol, has never been produced in volume as a standard chemical industry product, leaving prospects for low-cost production unclear.[10]

Energy density is critical for volume- or weight-limited systems, military and civilian. Lower energy density means either bigger fuel tanks or else a sacrifice in range, tradeoffs that may be acceptable in passenger cars but not for military equipment or aircraft of any type (and volume tradeoffs become more severe in passenger cars downsized to save weight and increase fuel economy). Infrastructure, along with the limited tolerance of existing vehicles for ethanol content in gasoline, will also continue to limit civilian consumption and, because pipelines, pumps, and many storage facilities are incompatible with ethanol, it generally reaches fuel blenders by truck or railroad at transport costs exceeding those for petroleum.

Ethanol does replace some imported petroleum. Since ethanol's energy content is two-thirds that of gasoline by volume, replacing 10 percent of US gasoline with ethanol in effect displaces 6-7 percent of gasoline consumption (disregarding the effect of ethanol's lower energy content on fuel mileage). Imports make up less

[10] *Liquid Transportation Fuels from Coal and Biomass: Technological Status, Costs, and Environmental Impacts* (Washington, DC: National Academies Press, 2009), pp. 206-207.

than half of US petroleum consumption, so that corn ethanol displaces no more than 3-4 percent of imports. This is far less than US *exports* of petroleum products, which have recently fluctuated in the range of 20-30 percent of imports.[11]

[11] See the Energy Information Administration data series, covering the past two decades, for "Weekly U.S. Imports of Crude Oil and Petroleum Products," <www.eia.gov/dnav/pet/hist/LeafHandler.ashx?n=PET&s=WTTIMUS2&f=W> and "Weekly U.S. Exports of Crude Oil and Petroleum Products," <www.eia.gov/dnav/pet/hist/LeafHandler.ashx?n=PET&s=WTTEXUS2&f=W>.

Table 2

Fuel Properties [a]

	Density (lb/gal)	*Energy Density* Gravimetric (Btu/lb)[b]	Volumetric (Btu/gal)[b]
Ethanol	6.6	11,500	76,000
Butanol	6.75	14,800	100,000
Gasoline	6.0-6.5	18,000-19,000	114,000-116,000
Jet Fuel	6.5-7.1	18,400-18,600	124,000-126,000
Diesel Fuel[c]	7.1	18,000-19,000	128,000-130,000
Biodiesel	7.3	16,200	118,200

[a] Values given are representative. As explained in Appendix A2, fuel specifications, which cover many more properties than those listed above, take the form of minimums, maximums, or allowable ranges and actual properties may vary considerably.

[b] Btu, standing for British thermal unit, is the energy necessary to raise the temperature of one pound of water by one degree Fahrenheit.

[c] No. 2, as sold for road vehicles.

Sources: Department of Energy, Alternative Fuels Data Center, <www.afdc.energy.gov/>; Energy Information Administration, <www.eia.gov/oiaf/renewable.html>; *Biodiesel Handling and Use Guide*, Fourth Edition, NREL /TP-540-43672 (N.p.: National Renewable Energy Laboratory, December 2009). Butanol – M. Wu, M. Wang, J. Liu, and H. Huo, *Life-Cycle Assessment of Corn-Based Biobutanol as a Potential Transportation Fuel*, ANL/ESD/07-10 (Argonne, IL: Argonne National Laboratory, November 2007), Table 1, p. 6. Jet fuel - *World Fuel Sampling Program*, CRC Report No. 647 (Alpharetta, GA: Coordinating Research Council, June 2006); *Department of Defense Handbook: Aerospace Fuels Certification*, MIL-HDBK-510-1A(USAF) (N.p.: US Air Force, February 8, 2010).

4

DROP-IN FUELS FOR THE US MILITARY

Ethanol cannot replace hydrocarbon fuels. DoD has no interest and civilian consumption beyond RFS2 would mean a costly and protracted transition requiring changes to both vehicles and infrastructure. Drop-in synthetics hold far more promise in both military and civilian spheres. Such fuels will have to meet the same or very similar specifications as the petroleum fuels they supplement or replace. Biodiesel, for example, does not qualify as a drop-in fuel; renewable diesel, which can be mixed in any proportions with petroleum diesel, does.

Making Synthetics

Numerous pathways to drop-in synthetics have been explored and some have been commercialized. Natural gas, for instance, consists chiefly of methane, CH_4; gasoline or kerosene can be made by decomposing these molecules and then recombining the constituents in new configurations, minus some of the hydrogen, to yield longer chains of the general form C_xH_y. Similar processes can be used with biomass or coal as feedstock, beginning with gasification. The "synthesis gas" produced can then be processed into liquid hydrocarbons. Proven commercial processes exist and considerable experience has accumulated. Technical development in the chemical and petrochemical industry is never easy, scale-up in particular, but costs and financial risks have been the greater obstacles. Today, those costs include environmental costs. While environmental costs remain unpriced, no one any longer disregards them as during the periodic bursts of enthusiasm for making synthetic fuels from coal, typically stimulated by anxiety over apparently declining petroleum supplies, during the first half of the last century and on into the 1980s.

Germany, with little oil, produced synthetic gasoline from coal during World War II. In the middle 1950s, the South African enterprise Sasol began making synthetic fuels with one of the two methods originally developed in Germany, Fischer-Tropsch (FT) processing (Box C). South Africa's apartheid-era government, which feared embargoes on imported oil, heavily subsidized production and today Sasol makes synthetic gasoline, diesel fuel, and jet fuel from natural gas as well as coal, supplying around 30 percent of the country's transportation fuels.

In the United States, the Bureau of Mines, part of the Interior Department, explored synthetic fuels made from coal and shale beginning around the time of the First World War. With the 1973-74 Arab oil embargo and the establishment of the Energy Research and Development Administration in 1974, the portfolio passed to the new agency and in 1977, with the next reorganization of energy responsibilities, to the Department of Energy (DOE). From the 1950s, military agencies explored synthetics extensively, in search of high-density jet and rocket fuels that promised greater range and payload for aircraft and missiles (see Appendix A4). In the 1970s, concerned like others over possible fuel shortages, DoD began supporting work on synthetics made from domestic coal, shale, and tar sands as substitutes for ordinary kerosene.[1] As for recent work to certify aircraft and other equipment for routine operation on synthetics, R&D in the 1970s and 1980s assessed implications for the performance and longevity of powerplants originally designed to run on petroleum. Also in the 1970s the Air Force was evaluating the shift from JP-4, its previous standard, to JP-8 and the National Aeronautics and Space Administration was studying the contributions of aircraft to air pollution, work that included research on combustion phenomena in jet engines and gas turbines. Although synfuels funding dropped off in the latter part of the 1980s as oil prices returned to low levels, R&D continued on other parts of the agenda, especially emissions such as unburned hydrocarbons and smog-forming nitrogen oxides from jet engines and gas turbines. These several streams of R&D contributed to what is now a substantial technology base for synthetic jet fuels, drawn on in Appendixes A3 and A4.

Little more than a decade ago when the Defense Science Board called on the military to give greater attention to energy efficiency its report said nothing about alternative fuels.[2] As so many times in the past, new interest came with rising oil prices. In 2009

[1] For a summary, see A. V. Churchill, C. L. Delaney, and H. R. Lander, "Future Aviation Turbine Fuels," *Journal of Aircraft*, Vol. 15, No. 11, November 1978, pp. 731-734.

[2] *More Capable Warfighting through Reduced Fuel Burden* (Washington, DC: Office of the Under Secretary of Defense for Acquisition and Technology, May 2001).

Secretary of the Navy Ray Mabus announced an ambitious set of energy goals for the Navy and Marine Corps, including a "Great Green Fleet" powered by non-petroleum fuels to be deployed in 2016. Also in 2009 Congress established, as part of that year's Defense Authorization Act, a Pentagon directorate for Operational Energy Plans and Programs. And in mid-2012 the Navy joined with DOE and the US Department of Agriculture (USDA) in a $510 million program to support construction of biorefineries to supply military fuels.

Much as EISA carved out new markets for biofuels in road transportation, DoD and the services promise new markets for synthetic jet fuel. According to the Pentagon's 2012 *Operational Energy Strategy,* the Air Force expects "to be ready to cost-competitively acquire 50 percent of its domestic aviation fuel requirement, or roughly 387 million gallons of JP-8 per year, via an alternative fuel source by 2016." For the Navy, the "2020 goal is to use alternative sources for half of all energy consumption afloat, which will require 300 million gallons of biofuels." The Army, for its part, "does not have specific goals or policies in place for procurement or widespread adoption of alternative fuels."[3]

That the three services adopted markedly different stances should be no surprise, considering that the Air Force and Navy have never been able to agree on a common jet fuel, which if nothing else would make logistics and supply chain management a bit easier. The Air Force has built escape hatches into its alternative fuels planning, conditioning purchases on price parity

[3] All quotations from *Operational Energy Strategy: Implementation Plan* (Washington, DC: Department of Defense, March 2012), p. 17.

with petroleum, leaving open the possibility of using synthetics made from fossil sources such as natural gas, and stating its intent simply to be "ready" to buy alternative fuels. The Navy, its vision of a Great Green Fleet the public face of DoD fuels policy, says nothing about pricing and commits itself to biofuels. Although these and other statements from the armed forces are hardly binding, the overall policy is plain enough: DoD expects, or at least hopes, that the prospect of sales to the military will induce private investment.

To strengthen the incentives, each of the participants in the three-year Navy/DOE/USDA "Advanced Drop-In Bio Fuels Production Project" plans to contribute $170 million for plant design and construction (or retrofits) to be cost-shared with industry.[4] The Defense Production Act (DPA) provides the framework for Navy and DOE participation. Dating from the time of the Korean War, the DPA authorizes the president to "provide incentives to domestic manufacturers to create, maintain, protect, expand, or restore" capacity for producing critical defense matériel.[5] DPA authority does not extend to USDA; its contributions, focused on feedstock, will be channeled through the Commodity Credit Corporation.

[4] "Memorandum of Understanding between the Department of the Navy and the Department of Energy and the Department of Agriculture," <www.usda.gov/wps/portal/usda/usdahome?contentidonly=true&contentid=renewable_energy.xml>.

[5] "Funding Opportunity Announcement: Defense Production Act Title III Advanced Drop-In Biofuel Production Project," Air Force Research Laboratory, <www.fbo.gov/index?_cview=0&id=e95eb37a21ff0be1eeb7679515490e3d&mode=form&s=opportunity&tab=core >, p. 1. The Air Force administers DPA initiatives for all of DoD; it is not a partner in the biofuels project.

Available fiscal 2012 funds provided $30 million for Phase 1 awards supporting preliminary design, which could include some R&D; DoD announced three awards totaling $16 million in May 2013. After a "down-select," Phase 2 funds will support "construction, commissioning and performance testing" of up to three plants. Proposal guidelines call for capacity of at least 10 million gallons per year of fuels that meet military specifications for JP-5, JP-8, or F-76 (Navy bunker fuel) at costs competitive with jet fuel priced at wholesale. Industrial participants must put up half or more of expenditures. In their fiscal 2013 budget requests, DOE seeks $100 million and the Navy $70 million to fund Phase 2 proposals; USDA has sought its full commitment, $171 million.[6]

Thirty million gallons per year from the DPA project, less than 1 percent of DLA Energy's annual purchases, would not get the Navy and Air Force very far toward fulfilling their consumption goals, and no matter industry's response there seems unlikely to be much biokerosene for the services to buy in 2016. A working group of the intergovernmental US-Canadian North American Technology and Industrial Base Organization reported in 2010 that "At present there is not an existing domestic industrial base or favorable business climate to support a rapid migration to use of alternative fuels sources for military aviation systems."[7] A study released

[6] Anthony Andrews, Kelsi Bracmort, Jared T. Brown, and Daniel H. Else, *The Navy Biofuel Initiative Under the Defense Production Act,* R42568 (Washington, DC: Congressional Research Service, June 22, 2012), p. 13.

[7] *Industrial Base Assessment of Alternative Fuels for Military Use* (Dayton, OH: Universal Technology Corporation for NATIBO, June 7, 2010), p. xxiii. Officials from the defense departments of Canada and the United States serve on NATIBO.

in 2011 by DoD itself concluded that service goals for consumption of renewable fuels could probably not be attained, and that if they were to be reached DoD's annual fuel costs might rise by more than $2 billion.[8] A recent industry survey projects biojet capacity of no more than 150 million gallons in 2015, and perhaps as little as 25 million gallons.[9]

Although the Air Force and Navy have publicized their alternative fuels plans at every opportunity, industry's response seems tepid. Investors recognize that DoD's alternative fuels policies lack legislative underpinnings of the sort created by EISA for road fuels and that full funding by Congress for the DPA project cannot be assured. Nor has the much larger market for commercial jet fuel attracted strong interest. Potential suppliers know that airlines, competing fiercely among themselves in a low-margin industry that has seen frequent bankruptcies, will not pay a premium for biofuels except perhaps in small quantities to support green advertising campaigns. While several companies have recently announced plans to produce synjet from natural gas or biomass, which if any of these will come

[8] "DoD would require more than 40 percent of the total projected U.S. drop-in renewable fuel supply (regardless of fuel type) in 2020, just to meet the military Services' stated demand for 745 million gallons." *Opportunities for DoD Use of Renewable and Alternative Fuels* (Washington, DC: Department of Defense, July 2011), p. iii. Congress requested this study in the 2010 Defense Authorization Act.

[9] Mary Solecki, Anisa Dougherty, and Bob Epstein, *Advanced Biofuel Market Report 2012: Meeting U.S. Fuel Standards* (San Francisco: Environmental Entrepreneurs, n.d. [appears to be September 2012]), p. 6. Some provisions in US law treat fuel production in Canada as if it were domestic and this survey aggregates biofuels capacity in the two countries.

to fruition, or be delayed or canceled (as for so many proposed cellulosic ethanol plants), will not be known for several years.[10]

Box C

Synthetic Fuels from Coal and Other Feedstocks

A great many routes lead from biomass to liquid hydrocarbon fuels. Synthesis gas, or syngas, a mixture predominately of hydrogen and carbon monoxide, can be made from biomass of any sort, just as it can be made from natural gas or coal (hundreds of coal gasification plants operate around the world and many US cities piped coal gas, or town gas, to homes and businesses before natural gas reached the market in volume after World War II). The gas can then be made into liquid hydrocarbon fuels. There are also direct liquefaction routes to synthetic gasoline and kerosene and processes with alcohols as intermediate products.

Germany lacks oil and during World War II made synthetic fuels from coal using two quite different methods, Bergius hydrogenation and Fischer-Tropsch (FT) processing.[a] By 1944 German firms were operating 20 plants, split almost evenly between the two methods, with annual capacity approaching 1.5 billion gallons. This might seem an impressive figure; in fact, it could not keep the Nazi war machine going and today would amount to four days supply for American motorists.

Coal, not itself a hydrocarbon, consists mostly of large, heavy molecules (well over 100 carbon atoms typically, almost as many hydrogen atoms, plus other elements). To make liquid fuels, these molecules must be broken down and the non-hydrocarbon constituents removed. Bergius and FT plants do this differently and produce product streams

[10] John Morris Berlin, "Fueling Progress," *Aviation Week & Space Technology*, September 17, 2012, p. 42.

with distinct chemistries. The Bergius process yields many ring-like hydrocarbons, good for making high-octane aviation gasoline. Depending on choice of reaction temperatures and catalysts, FT processing can make a range of products, including gasoline.[b] The usual choice, for reasons of process efficiency, has been long-chain paraffins, synthetic paraffinic kerosene, appropriate for diesel or jet fuel. (One of the main hydrocarbon classes, paraffin molecules take the form of linear or branched chains with carbon atoms joined exclusively by single or "saturated" bonds—see Appendix A1.)

Because of the environmental costs, synthetic fuels from coal no longer excite much interest.[c] Containing more carbon than petroleum and much more than natural gas, coal releases more CO_2 per unit of useful energy than other fossil fuels and is also the worst offender in terms of toxic emissions (sulfur, mercury), worker deaths and injuries, and, with the possible exception of oil sands, degradation of terrestrial environments. The 2007 Energy Independence and Security Act bars federal agencies from purchasing alternative fuels with greater life-cycle greenhouse emissions than petroleum, including effects such as indirect land use changes. So long as this provision remains part of US law, the Defense Department can buy coal-based synfuels only for R&D and testing.

Shell, the only firm besides Sasol to have built significant FT synfuels capacity, opened a small plant fed by natural gas in Malaysia in 1993 and a much larger complex in Qatar in 2011. The Malaysian plant, after rebuilding and expansion in the late 1990s, represents a reported investment of around $1 billion for a capacity of 225 million gallons of diesel fuel per year. In Qatar, Shell expects to produce over 2 billion gallons of diesel fuel annually and a nearly equal volume of other, nonfuel products.[d] Shell is now evaluating a comparably sized US facility.[e] If built, it would be sited in Louisiana, for easy access to shale gas. In December 2012 Sasol announced its own plans to invest in Louisiana. The company has said it will spend more than $10 billion on a GTL plant capable of making 1.5 billion gallons per year of diesel fuel.[f]

Methane-to-methanol, finally, is a standard industrial process and Mobil, now Exxon Mobil, commercialized methanol-to-gasoline (MTG) in the mid-1980s, based on proprietary catalysts, putting up a 220 million gallon per year plant in New Zealand fed by natural gas.[g] The country's government, which aimed to reduce dependence on imported oil, provided financing and retained three-quarters ownership. Gasoline production continued until 1997, when low oil prices led to a shift to other end products.

[a] Raymond G. Stokes, "The Oil Industry in Nazi Germany, 1936-1945," *Business History Review*, Vol. 50, No. 2, Summer 1985, pp. 254-277. Friedrich Bergius developed the method that bears his name (and won him the 1931 Nobel prize in chemistry) at the time of the First World War; the Fischer-Tropsch process dates from the 1920s. For work in the United States, see Anthony N. Stranges, "The US Bureau of Mines' Synthetic Fuel Programme, 1920-1950s: German Connections and American Advances," *Annals of Science*, Vol. 54, 1997, pp. 29-68. On earlier US policies, see Richard H.K. Vietor, "The Synthetic Liquid Fuels Program: Energy Politics in the Truman Era," *Business History Review*, Vol. 54, No. 1, Spring 1980, pp. 1-34.

[b] John J. Marano and Jared P. Ciferno, *Life-Cycle Greenhouse-Gas Emissions Inventory For Fischer-Tropsch Fuels* (N.p.: Energy and Environmental Solutions, LLC for Department of Energy, June 2001), p. 14.

[c] That said, a National Research Council panel has reported somewhat positively on production of drop-in fuels from feedstock streams mixing biomass with coal, at least if carbon capture and storage were to be incorporated. *Liquid Transportation Fuels from Coal and Biomass: Technological Status, Costs, and Environmental Impacts* (Washington, DC: National Academies Press, 2009).

[d] Andrew Callus and Peg Mckey, "Shell Ramps Up Qatari Gas-to-Diesel Money Machine," Reuters, August 28, 2012, <www.reuters.com/article/2012/08/28/shell-gtl-idUSL6E8JP1F420120828>.

[e] Russell Gold, "Shell Weighs Gas-to-Diesel Processing Facility for Louisiana," *Wall Street Journal*, April 4, 2012.

[f] Clifford Krauss, "South African Company to Build US Plant

to Convert Gas to Liquid Fuels," *New York Times*, December 3, 2012.

[g] Xinjin Zhao, Ron D. McGihon, and Samuel A. Taback, "Coal to Clean Gasoline," *Hydrocarbon Engineering*, September 2008. As this article's title implies, Exxon Mobil has promoted its MTG process mostly for converting gasified coal to vehicle fuel. On methanol-to-gasoline more generally, see, e.g., Steven D. Phillips, Joan K. Tarud, Mary J. Biddy, and Abhijit Dutta, *Gasoline from Wood via Integrated Gasification, Synthesis, and Methanol-to-Gasoline Technologies*, NREL/TP510047594 (Golden, CO: National Renewable Energy Laboratory, January 2011), especially pp. 4-8.

Synthetics as Substitutes for Petroleum

In addition to meeting specifications similar though not necessarily identical to those for petroleum, a synthetic will have to "demonstrate that it is 'fit-for-purpose' as jet fuel by having other defined properties and characteristics that fall within the range of experience with conventional, petroleum-derived jet fuel."[11] In 1999, after satisfying itself on such matters, the United Kingdom's Ministry of Defence (MOD), keeper of the Jet A-1 standard, certified Sasol's synthetic paraffinic kerosene (SPK) blended at up to 50 percent with petroleum.[12] Approvals by other national bodies

[11] Clifford A. Moses, *Development of the Protocol for Acceptance of Synthetic Fuels Under Commercial Specification* (San Antonio: Southwest Research Institute, September 2007), p. iii. A list of DoD fitness-for-purpose tests appears on p. 311 of *Department of Defense Handbook: Aerospace Fuels Certification*, MIL-HDBK-510-1A(USAF) (N.p.: US Air Force, February 8, 2010).

[12] Paul Morgan, "An Overview of Sasol's Jet Fuel Journey," presentation at 20th World Petroleum Congress, Doha, Qatar, December 4-8, 2011.

followed, and international flights landing in South Africa could thereafter fill their tanks routinely with Sasol's blend, sometimes termed semisynthetic. A few years later the US Air Force began tests on semisynthetics, beginning with the venerable B-52 bomber (see Appendix A2), and in 2007 established an Alternative Fuel Certification Office (AFCO) to oversee such testing. By the end of 2011 AFCO had certified all of the service's aircraft types on blends containing up to 50 percent SPK made from coal or natural gas.

The 50 percent limitation stems from the absence of ring-shaped aromatic hydrocarbons in synthetics. Gasification and FT processing remove any and all such molecules from the original feedstock, breaking them apart to produce an output stream consisting entirely of chain-like paraffins. As explained in Appendix A3, fuels low in aromatics can cause seals and gaskets to leak, especially in older equipment. To avoid this, semisynthetic jet fuel specifications as established first by MOD and later by DoD require a minimum aromatic content of 8 percent. With crude oils containing a high percentage of naturally occurring aromatics and many of these remaining in refined fuels, 50 percent or more petroleum mixed with SPK provides sufficient aromatic content to protect against dangerous (and polluting) leakage. (Specifications for common petroleum fuels do not set minimums for aromatics, only maximums, since high percentages of aromatics have undesirable consequences including exhaust smoke, as also discussed in Appendix A3.) Sasol's 100 percent synthetic jet fuel, certified by MOD and ASTM in 2008, carries a package of additives to compensate for the lack of aromatics.[13] AFCO testing with full synthetics was

[13] Sasol spent seven years, 1991-1999, shepherding its FT semisynthetic through approvals and another seven gaining

underway but incomplete at the end of 2012; when DoD issues standards they will no doubt require special additives too.

SPK can also be made without gasification from animal fats and oil-rich energy crops such as camelina and algae. (Camelina has been a popular candidate for biofuels production because it promises relatively high yields over wide-ranging conditions of sun and water on land considered marginal or unsuitable for food crops.) Production begins much as for renewable diesel, with a hydroprocessing/hydrotreating stage that adds hydrogen and breaks down large molecules into the kerosene range.[14] These fuels are commonly referred to as HEFA, for hydroprocessed esters and fatty acids, or HRJ, for hydroprocessed renewable jet fuel. Along with FT blends, DoD has tested semisynthetics labeled HRJ-5 and HRJ-8 and DLA Energy has purchased renewable diesel for testing in ground vehicles and small quantities of algae-based jet and diesel fuel, also for R&D and

acceptance for its 100 percent SPK fuel. Similar synthetics from other suppliers are now eligible for expedited approvals by MOD, ASTM, and other authorities. The Cooperative Aviation Alternative Fuels Initiative, CAAFI, established in 2006 by airlines, engine manufacturers, and several government organizations, including the US Federal Aviation Administration, makes available roadmaps for approval of commercial synjet fuels; they that can be found at <www.caafi.org/>.

[14] For a recent review, see Juan Carlos Serrano-Ruiz, Enrique V. Ramos-Fernández, and Antonio Sepúlveda-Escribano, "From Biodiesel and Bioethanol to Liquid Hydrocarbon Fuels: New Hydrotreating and Advanced Microbial Technologies," *Energy & Environmental Science*, Vol. 5, 2012, pp. 5638-5652.

testing purposes.[15]

The technology base for synthetic jet fuels built up since the 1970s has continuing relevance for evaluating newer generations of alternatives. FT and HRJ kerosenes exhibit minor advantages over petroleum because they burn more cleanly and minor disadvantages because of potential seal leakage and slightly lower energy density. The review in Appendixes A3 and A4 finds nothing that seems significant enough to argue either for or against DoD adoption of synthetics based on technical characteristics.

Technological immaturity slowed the pace for cellulosic ethanol. This is less of a concern for synthetic jet fuel. SPK process technology seems well in hand. The questions concern costs, greenhouse impacts, and investors' judgments of risks and opportunities ahead.

[15] *Opportunities for DoD Use of Renewable and Alternative Fuels* (Washington, DC: Department of Defense, July 2011), Table 2-3, p. 2-4.

5

COMPETING WITH PETROLEUM

Range Fuels, one of the more heavily publicized cellulosic ethanol startups, closed its doors at the end of 2011. The company had the backing of prominent Silicon Valley venture capitalist Vinod Khosla and benefited from over $150 million in loan guarantees and grants from DOE and USDA.[1] In September 2012 Terrabon, known for its proprietary alcohol-to-gasoline process and as a subcontractor on biojet R&D work funded by the Defense Advanced Research Projects Agency (DARPA), filed for bankruptcy after failing to attract next-round financing.[2] These signs of fragility in the

[1] Mario Parker, "Range Fuels Cellulosic Ethanol Plant Fails, US Pulls Plug," December 2, 2011, *Bloomberg.com*, <www.bloomberg.com/news/2011-12-02/range-fuels-cellulosic-ethanol-plant-fails-as-u-s-pulls-plug.html>. Government financial support for Range Fuels came during the George W. Bush administration. The firm's plant, designed to produce ethanol from wood chips, was sold in a bankruptcy proceeding early in 2012.

[2] "Alcohol to Jet Fuel: One Step Closer after Terrabon Begins R&D Program," *Biofuels Digest*, July 26, 2011, online;

nascent biofuels industry coexist with evidence of vitality. Shortly after the Terrabon filing, Joule, a small US company, announced a strategic partnership with the German automaker Audi, part of the huge Volkswagen group. Under the agreement the two firms expect to cooperate on further development of Joule's one-step process for making ethanol or biodiesel from patented strains of bioengineered cyanobacteria.[3]

The troubles encountered with cellulosic ethanol— in October 2012 BP canceled construction of a plant in Florida that had been announced four years earlier— contrast with the ability of corn ethanol suppliers to quickly ramp up production in response to RFS2.[4] Corn ethanol process know-how was readily available, capital costs were low, and feedstock available for purchase in an established market. Productivity in growing corn benefited from decades of experience with hybrid seeds, agrochemicals, and mechanization. Both farmers and their customers operate on industrial scale. Nothing similar exists for dedicated energy crops, much less agricultural residues.[5] Government policies rewarded

"Advanced Biofuels Pioneer Terrabon Files for Chapter 7 Bankruptcy: One-Off or Trend?" *Biofuels Digest*, September 10, 2012, online.

3 Kevin Bullis, "Audi Backs a Biofuels Startup," *Technology Review*, September 17, 2012. Sometimes termed blue-green algae, cyanobacteria contain chlorophyll; they are prokaryotes whereas algae are eukaryotes.

4 Christopher Martin and Mario Parker, "BP Cancels Florida Ethanol Plant, Extends Renewable Exit," *News from Bloomberg: Bloomberg Businessweek*, October 25, 2012, <www.businessweek.com/news/2012-10-25/bp-cancels-plants-to-build-florida-cellulosic-ethanol-plant>.

5 For an account of disappointments with cultivation of jatropha, see Carol Matlack, "A Green Fuel Effort Awash in

corn ethanol producers that took advantage of readily available opportunities: there was little to learn and money to be made. Biodiesel stumbled by comparison because of weaker policy incentives relative to profit-and-loss calculations.

Market Behavior and Investment Dynamics

A company able to produce JP-10, a high energy-density synthetic that DLA Energy buys for cruise missiles and resells to the services at $20 per gallon (Appendix A4), may enjoy some leverage over pricing, depending on how many other suppliers could fill DoD's orders. Leaving aside such exceptions, fuels are commodities, branded for marketing purposes but otherwise undifferentiated and selling at common prices. While the larger state-owned oil companies command market power, no single producer, not even Saudi Arabia, can set prices worldwide. Nor can the Organization of Petroleum Exporting Countries, a shaky cartel that from its beginnings has struggled to reach agreements and hold members to them. Producers of generic alternative fuels—HRJ-8 or synthetic Jet A-1—will have to accept prices set in the global marketplace, absent policy interventions. Unlike innovators in many other industries, they have no way to differentiate themselves through unique products offering features otherwise unavailable.

Innovation will occur, bringing costs of biofuels down. Public and private R&D groups are working on bioengineered organisms and novel production processes that bypass the costly, complex, and energy-

Red Ink," *Bloomberg Businessweek*, March 19-March 25, 2012, pp. 68-69.

intensive steps in making FT or HEFA hydrocarbons. But no one can know how rapidly innovation will proceed, at what levels costs will settle, or whether energy inputs can be reduced to the point that GHG emissions will be significantly less than for petroleum. And prices for biomass feedstocks could rise, as they have for corn and soybeans, turning a once-profitable biorefinery into a money-losing operation. Or prices for oil could fall, with the same outcome.

While no one can predict oil prices with assurance, by most accounts they will remain high. In its 2013 *Annual Energy Outlook*, the EIA raised projected oil price trends over the next several decades above those in the 2012 edition of this annual publication.[6] Prices will probably also remain volatile. They respond sensitively to global economic and geopolitical events, rising with war or rumors of war, especially if oil-producing regions or shipping routes such as the Strait of Hormuz seem threatened, and falling in economic recessions. Even small unanticipated shocks can cause large price swings (i.e., relative to the magnitude of the change in supply or

[6] *Annual Energy Outlook 2013,* DOE/EIA-0383(2013) (Washington, DC: Energy Information Administration, April, 2012), Fig. 49, p. 58; *Annual Energy Outlook 2012*, DOE/EIA-0383(2012) (Washington, DC: Energy Information Administration, June 2012), Fig. 64, p. 72. EIA increased prices in all three of its projected scenarios, High Oil Price, Reference, and Low Oil Price. The intermediate Reference case is defined as "a business-as-usual trend estimate, given known technology and technological and demographic trends." See *Annual Energy Outlook 2013*, p. ii. The EIA scenarios should not be considered forecasts; they are better viewed as self-consistent time series associated with EIA's large and complex economic models.

demand).[7] Prices could decline for reasons including: an end to the embargo on oil exports from Iran, leading to increased supplies on the world market and relaxed upward pressures on prices (the more so should Iran, perhaps after a change in government, welcome inward investments by multinational energy firms); political stability in Iraq likewise coupled with policies designed to attract foreign direct investment; post-Chavez policy shifts in Venezuela; government reforms that raised production in Nigeria; new discoveries elsewhere comparable to those in the waters off Brazil.

The sheer size of the global energy business sets petrochemicals off from other industries. Shell tops the global Fortune 500 and Exxon Mobil led the list of US-based firms in 2012 and trailed only Wal-Mart in 2013. Big oil companies have enormous cash reserves, much flexibility in financing, and considerable influence over government policies, the more so if state-owned. They expect to prosper almost regardless of price levels: market fluctuations that would bankrupt a biofuels startup in months may barely register in their financial reports. A further difference sets the risks faced by biofuels firms apart from those in other commodity markets: the *price* commanded by crude oil greatly exceeds the *cost* of producing much of that oil. These costs, including exploration, drilling, and pumping, vary much more than the costs of producing commodities such as wheat or aluminum. Some oil costs as little as $1 per barrel to extract, leaving aside royalties. As Table 3 indicates, costs below $20 per barrel are not uncommon. Prices as high as they were in 2012, an average of about $100 per barrel, generate mountains of cash for

[7] James L. Smith, "World Oil: Market or Mayhem?" *Journal of Economic Perspectives*, Vol. 23, No. 3, Summer 2009, pp. 145-164.

enterprises fortunate enough to have access to low-cost reservoirs.

Table 3

Crude Oil Production Costs

Region	*Typical Production Cost* (dollars per barrel)[a]
Saudi Arabia, Iran, Kuwait	1-2
Nigeria	4
Venezuela	5-6
North Sea	8
Russia	8-20
Gulf of Mexico	10

[a] Including exploration and drilling

Source: Roberto F. Aguilera, et al., "Depletion and the Future Availability of Petroleum Resources," *Energy Journal*, Vol. 30, No. 1, 2009, pp. 141-174.

Costs rise steeply in locations such as the Arctic or deep offshore waters and petrochemical firms pace their investments depending on market dynamics. Anticipating sustained high prices, they may drill under difficult conditions in places where costs will never approach those in Table 3. In recent years, exploratory drilling and production in the United States increased in direct response to rising prices. Drilling is cheap in most of the lower 48 states; companies started twice as many new wells in 2010 (over 16,000) as in 2000 (8100).[8] If

[8] Drilling costs vary widely but seldom exceed a few million dollars. "Costs of Crude Oil and Natural Gas Wells Drilled," Energy Information Administration, September 7, 2012, <www.eia.gov/dnav/pet/pet_crd_wellcost_s1_a.htm>. For numbers of well drilled, see *Annual Energy Review 2011*,

demand slackens, exploration slows, drillers begin fewer wells, and high-cost operations may shut down temporarily. Oil in the ground will still be there next year or next decade and leases can be bought, sold, or renewed.

Biofuels firms must also weigh the risks of strategic behavior by rivals. Major petrochemical companies including BP and Shell have multibillion-dollar investments in biofuels.[9] For an oil producer, a stake in biofuels may serve as a hedge, much like the R&D many of these companies conducted decades ago on synfuels from coal, or part of a diversification strategy. Fossil fuel firms with a sideline in biofuels could if they wished sell alternative fuels indefinitely at prices below their costs (just as they sometimes sell oil from harder-to-access, expensive fields at below cost), to the extent costs can be allocated in meaningful fashion in diversified organizations, justifying such behavior as needed to move down learning curves even as their practices disadvantage rivals. While some forms of predatory economic behavior are illegal under US law, they do occur (and are nearly impossible to prove); in other parts of the world, formal restraints may be nonexistent or unenforced. Big companies could also choose to buy up biofuels startups. Normal business behavior in many industries, including pharmaceuticals and information technology as well as energy, has been to seek growth by bringing in innovations from the outside, often through acquisitions. In so doing, the purchaser sometimes stumbles, alienating key technical personnel

DOE/EIA-0384(2011) (Washington, DC: Energy Information Administration, September 2012), Table 4.5, p. 97.

[9] See, e.g., Ken Wells, "Focus on Energy: Green Giants?" *Bloomberg Businessweek*, May 14-May 20, 2012, pp. 57-58.

in the acquired firm, those who understand the technology and its intended applications, or faltering because of conflicting intra-organizational interests. Even if the *intent* were to integrate a biofuels acquisition into ongoing operations, or to enter a new line of business, the *outcome* may be to bury a promising innovation. Sometimes, of course, that outcome may be intended.

Costs and Scale

Like startups in many industries, biofuels firms expect to lose money for a time, with profits following as revenues mount, R&D and capital expenditures decline, and plant-level productivity rises. Petroleum companies have been driving down costs in this fashion for a century. They have also been able to tap scale economies for oil that may never be attainable for biofuels. The bottlenecks lie in collecting and transporting feedstock, not in processing. Most biomass is low in value and costly to transport relative to value. Shipping Saudi Arabian crude to North America costs little compared to its value; oil can be pumped and piped, transported by supertanker, barge, or railroad tank car. Energy crops or agricultural residue, waste by another name, must first be harvested or collected and then trucked (perhaps moved by rail) to the biorefinery (or to a temporary storage facility, for seasonally available feedstocks such as corn stover). The bigger the biorefinery, the bigger the area from which it will have to draw feedstock. Transportation charges mount quickly toward half or more of the price the biorefiner pays at the plant gate. Dense, higher-value feedstocks, perhaps at some point algae, may be worth shipping greater distances. Regardless, the calculation has the same form: biorefineries will be located to balance delivered costs of biomass against scale economies in constructing, equipping, and operating the plant.

For oil refineries and biorefineries alike, bigger plants cost less to build and operate per unit of output. Scale economies may diminish or disappear past some size, but biofuels firms will probably never get anywhere near this point. The average yearly capacity of the 144 oil refineries in the United States approaches 2 billion gallons.[10] Today a 100 million gallon biorefinery, one-twentieth the size of the average oil refinery, counts as large. Corn ethanol plants average around 70 million gallons in annual capacity and biodiesel plants still less, 20 million gallons.[11] Proposal guidelines for the Navy/DOE/USDA program set minimum plant capacity at just 10 million gallons per year.

Late in 2012 DuPont began construction of a $200 million cellulosic ethanol plant in Iowa, at 30 million gallons per year said to be the world's biggest.[12] Number 72 on the (US) Fortune 500 list, DuPont could easily finance a larger facility. The firm's decision-makers perhaps view both the technology and the business plan as unproven. Yet even if they were fully confident, collection and aggregation of corn stover, their feedstock of choice, would limit scale.

Second-generation biorefineries will probably be centered within a radius of a few hundred miles of

[10] *Refinery Capacity Report* (Washington, DC: Energy Information Administration, August 2012), Table 1, p. 1. Not all these refineries are active; capacity exceeds demand.

[11] *Biorefinery Locations* < www.ethanolrfa.org/bio-refinery-locations/>; *Monthly Biodiesel Production Report* (Washington, DC: Energy Information Administration, August 2012).

[12] "DuPont Breaks Ground on World's Largest Cellulosic Biorefinery," *Biofuels Digest*, November 30, 2012, online.

feedstock supplies; industry analysts tend to believe that capacities much above 100 million gallons per year will be practical only for unusually dense feedstocks or in conjunction with liquefaction or gasification facilities located close to biomass sources, to take advantage of available scale economies in downstream if not upstream processing stages.[13] With capital investments several times those for corn ethanol plants and limited opportunities for scale economies, firms with proprietary processing technologies for cellulosic biomass might well decide to pursue businesses such as high-margin specialty chemicals in preference to commodity fuels.

Scale in petroleum processing increased as firms learned to design, build, and operate ever-larger refineries filled with ever-larger equipment. Companies often underestimated technical difficulties—no surprise, since creative engineers, entrepreneurs, and innovators are optimists almost by definition, while in executive ranks as well, organizations rarely promote pessimists to lofty positions. Then too, new processes and plants in the chemical and petrochemical industries carry greater technical (and cost) uncertainties than in most other manufacturing sectors. The complexity of chemical reaction sequences and the effects of catalysts, scale-dependent factors such as surface-to-volume ratios and laminar-to-turbulent flow transitions, with their effects on heat transfer and reaction rates, underlie many of the special difficulties.

The petrochemicals industry has been a technological leader, with a lengthy record of innovation by oil companies and also by smaller, specialized suppliers of

[13] Tom L. Richard, "Challenges in Scaling Up Biofuels Infrastructure," *Science*, Vol. 329, August 13, 2010, pp. 793-796.

hardware, knowledge, and services (e.g., Schlumberger in geophysical exploration, Universal Oil Products in refinery processes). Yet petrochemical firms, including those with extensive experience, abundant internal technical expertise, and the ability to identify and hire the best engineering and design specialists, construction contractors, and equipment suppliers in the world, err frequently in estimating plant construction costs and performance measures (throughputs, yields).[14] At 225 million gallons, Shell's 1990s-era FT plant in Malaysia dwarfs DuPont's biorefinery and is itself dwarfed by Shell's new GTL complex in Qatar. The Qatar plant employs FT technology similar to that in Malaysia, yet Shell reportedly spent nearly four times its original estimate of $5 billion on the Qatari facility, something like $19 billion, and the new plant has had a rocky startup, with unexpected breakdowns and high maintenance costs.[15]

Biofuels firms can expect similar difficulties in moving from pilot scale to production, even if plant size never climbs as it has in the petrochemical industry. With thin technical resources and limited financial cushioning they must attempt to commercialize and

[14] Edward W. Merrow, Kenneth E. Phillips, and Christopher W. Myers, *Understanding Cost Growth and Performance Shortfalls in Pioneer Process Plants*, R-2569-DOE (Santa Monica, CA: RAND, September 1981). This study, based on an examination of projects and practices in 34 oil and chemical companies, remains analytically relevant despite three decades of technological advances. Computerized process modeling and computer-aided plant design notwithstanding, the fundamentals of what chemical engineers call unit operations have not changed much.

[15] John M. Broder and Clifford Krauss, "A Big, and Risky, Energy Bet," *New York Times*, December 17, 2012.

scale up technologies for which little experience has accumulated. In this light, the disappointments encountered over the past half-dozen years by firms working on cellulosic ethanol should not seem surprising.

Fossil fuel producers have further advantages over suppliers of biofuels. Coal companies and oil companies with their employees and customers represent massive vested interests. Coal and oil have wrestled over favors from government for a century, whether subsidies, friendly enforcement of health, safety, and labor regulations, or trade protection. The biofuels industry is too new and small to compete on equal terms, except perhaps in alliance with agribusiness. Congress wrote the Renewable Fuel Standard into US law. Despite continuing opposition by some in Washington to policies linked with global warning, RFS2 seems unlikely to disappear entirely. Corn ethanol, notably, enjoys strong backing. Yet managers and investors in biofuels companies cannot disregard scenarios in which policymakers scale back or strike mandates for advanced biofuels.

6

ALTERNATIVE FUELS FOR THE DEPARTMENT OF DEFENSE

Energy innovation has much to offer DoD. The services would benefit from diesel-powered generators with greater fuel-efficiency, advances in soldier-portable rechargeable batteries for electrical and electronic gear, lighter and more efficient solar panels to power forward bases, and operational practices that save fuel without compromising mission requirements.[1] As often noted, truck convoys in Iraq and Afghanistan carrying fuel to feed diesel generators drew frequent attacks. Less fuel needed in remote areas would translate into fewer casualties. Fuel burden could be reduced in several ways, including more efficient generators. But changing the fuel would make no difference unless energy density could be increased substantially, which it cannot (see Appendix A4). Fuel might also be manufactured in the field from waste. This is technically possible. Such fuel

[1] *Energy Innovation at the Department of Defense: Assessing the Opportunities* (Washington, DC: Consortium for Science, Policy and Outcomes, March 2012), especially pp. 10-13 and 26-31.

could not be assumed to meet military standards for JP-8. Petroleum refineries rely on well-developed quality control practices, such as statistical process control, which troops in remote areas could hardly replicate, much less the laboratory tests that back them up. Since, for example, 10 parts per million of sulfur compounds can affect a fuel's lubricity, a measure of the ability to protect moving parts against wear, field-brewed kerosene might suffice for water heaters and cook stoves, but the Army would not want to run costly generators on untested fuel and no helicopter pilot is going to take off without knowing his or her tanks carry milspec JP-8 or its synthetic equivalent.

Access to Fuel

The armed forces will always have first call on petroleum supplies, certainly in times of war. As the Defense Science Board put the matter over a decade ago, "The issue is not whether DoD will be able to obtain the oil it needs to provide for our national defense, because it will."[2] The Strategic Petroleum Reserve holds some 30 billion gallons of crude oil.[3] As Table 1 showed, DoD bought 4.6 billion gallons of fuel in fiscal 2012, 3.5 billion gallons of this jet fuel. Refineries can turn only a fraction of each barrel of crude into kerosene; still, in 2012 US

[2] *More Capable Warfighting through Reduced Fuel Burden* (Washington, DC: Office of the Under Secretary of Defense for Acquisition and Technology, May 2001), p. 7.

[3] *Strategic Petroleum Reserve - Profile* (Washington, DC: Department of Energy, Fossil Energy, October 16, 2012), <www.fossil.energy.gov/programs/reserves/spr/index.html>.

refineries produced 22.5 billion gallons of jet fuel.[4] Airlines bought most of this. DoD's 3.5 billion gallons amounted to a bit more than 15 percent of domestic output, DLA Energy purchases much of its fuel overseas, and Air Force costs for jet fuel in 2012 ($8.8 billion) were comparable to those of the third largest US-based airline, American ($8.7 billion) and much below those of the biggest, United Continental ($13.1 billion).[5] For the US military, excepting some battlefield scenarios, fuel—and energy more broadly—is a management issue, not a strategic issue.

Pricing and Procurement

The figures above for airline fuel purchases are net of gains or losses on hedges, a form of derivative contract. Such contracts guarantee airlines (or railroads or electrical utilities) the right to buy a certain quantity of fuel at a certain price. Depending on the price paid for the contract itself, if oil prices rise above a set level the purchaser stands to save; otherwise the purchaser loses the sum paid for the contract. The intent is not to profit but to manage the risks of unexpected price jumps. By aligning actual expenditures on fuel more closely with budgeted expenditures, the enterprise reduces the probability that it will have to divert cash from other

[4] *Monthly Energy Review April 2013*, DOE/EIA-0035(2013/04) (Washington, DC: Energy Information Administration, April 25, 2013) Table 3.2, p. 39. Jet fuel made up a bit under 8 percent of US output of refined petroleum products in 2012.

[5] *Defense Logistics Agency Energy Fact Book Fiscal Year 2012* (Ft. Belvoir, VA: Defense Logistics Agency Energy, n.d.), pp. 30; *American Airlines Form 10-K, 2012*, online; *United Continental Holdings Form 10K, 2012*, online. Airline figures are calendar year.

planned outlays to pay for unexpectedly costly fuel. The services have complained repeatedly that "[by] continuing to rely on petroleum fuels, we are subject to the extreme price volatility in the global petroleum market and foreign supply disruptions."[6] Yet while DoD has legal authority to adopt hedging and other risk management practices common in the private sector, the Pentagon has not done so.

Instead, DLA Energy insulates its customers, the services, from price swings by keeping large cash reserves of its own; if drawn down during one fiscal year, the services pay more during the next budget cycle to replenish DLA's cash supply. In the past DLA sometimes had to seek supplemental appropriations from Congress to pay its bills; since 2006 the agency has adjusted the prices it charges the services at shorter intervals, limiting its need to request supplemental funds (while increasing the likelihood that one or more of the services will have to do so).

The Defense Business Board, an outside advisory body to the Pentagon, has recommended that DoD adopt price protection steps short of hedging, fearing congressional objections to such contracts and also expressing reservations because "DLA and senior DoD officials have very little experience" with complex financial management techniques such as derivative contracts.[7] The board also noted that DoD could if it

[6] Navy Captain James Goudreau, in Eric Watson, "Eye on Energy: How the US Navy Now Fuels the 'Tip of the Spear,'" *Watching World Energy Posts*, February 7, 2013, <www.watchingworldenergy.com/2013/02/07/eye-on-energy-fueling-the-tip-of-the-spear/>.

[7] *Re-examining Best Practices for DoD Fuel Acquisition*, Report FY11-06 (N.p.: Defense Business Board, 2011), p. 22.

wished contract with specialized financial services firms for advice on risk-adjusted fuel contracts, as private companies sometimes do, or indeed contract out the management of its fuel purchases.

To the extent that price fluctuations pose a genuine concern, Pentagon managers should begin by adopting standard business practices as recommended by the Defense Business Board. If and when substantial supplies of drop-in alternative fuels reach the market, they will sell at prices that track those for petroleum. At that point, DLA Energy will be just another customer. It could only insulate itself from price swings by adopting price protection methods of the sort previously avoided for petroleum, such as long-term fixed price contracts, or through DoD ownership of some production capacity, a step that, absent a true emergency such as that of World War II, would excite massive opposition.[8]

R&D and Production Capacity

DOE provides the bulk of federal R&D support for biofuels. The armed forces sponsor some R&D on fuels, but behave primarily as customers, as illustrated by the DPA project and certification of equipment on alternative fuels such as SPK blends. Certification is straightforward engineering, falling in category 6.4

[8] During World War II the federal government paid for some two-thirds of new capacity for the manufacture of military matériel. Paul A.C. Koistinen, *Arsenal of World War II: The Political Economy of American Warfare, 1940–1945* (Lawrence, KS: University Press of Kansas, 2004), p. 54. After the war's end, most of this plant and equipment was sold to private firms, although the government does still own a number of aircraft plants and shipyards, ammunition plants, and of course DOE's nuclear weapons facilities.

(Advanced Component Development and Prototypes) of DoD's budget system for R&D. Such work is not ordinarily a source of fundamentally new technical knowledge.[9]

DARPA's Biofuels Program, now completed, spent $100 million over the period 2007-2011.[10] Unlike DOE and the services, DARPA has no laboratories of its own; the Biofuels Program funded R&D carried out in industry, with some university participation. The intent was to develop commercially practicable technology for making jet fuel from biomass. At the conclusion of the program in 2011, DARPA management, evidently believing this objective had been accomplished, did not request funds for a follow-on, although DARPA's newer Living Foundries Program, which funds basic research in synthetic biology in search of "a revolutionary, biologically-based manufacturing platform" might at some point lead to biofuels innovations.[11]

[9] The Air Force has spent a cumulative total of about $125 million in recent years in testing aircraft and engines on alternative fuels. "DoD Energy Innovation Atlas," Consortium for Science, Policy and Outcomes, <www.cspo.org/projects/eisbu/atlas>.

[10] Before creating a formal program, DARPA supported some biofuels research at annual levels of a few million dollars through other programs. See *Department of Defense Fiscal Year (FY) 2013 President's Budget Submission: Defense Advanced Research Projects Agency Justification Book Volume 1* (Washington, DC: Department of Defense, February 2012), p. 160, and earlier DARPA budget justifications.

[11] *Department of Defense Fiscal Year (FY) 2014 President's Budget Submission: Defense Advanced Research Projects Agency Justification Book (Washington, DC: Department of Defense, April 2013)* pp. 42-43. For more on Living Foundries, see Elizabeth

With the end of major R&D support by DARPA, the DPA project becomes the military's chief effort to push forward with biofuels. Since neither bioethanol nor biodiesel appears viable without subsidies, it is hard to see how three 10 million gallon per year biorefineries, or even 100 million gallon per year biorefineries, could produce at costs competitive with wholesale jet fuel prices (at least without creative cost accounting) unless oil prices skyrocket within the next few years. Of course, if the cost competitiveness target proves unattainable DoD could choose to alter it. The Pentagon modifies contract terms frequently and sometimes accepts weapons systems that do not meet the original performance requirements. Major defense firms recognize these as standard practices. They know how to do business with the Pentagon, understand the risks, which on occasion have included termination of programs underway for many years (such as the Army's Comanche helicopter, cancelled when nearly ready for production after two decades of development), and the ins and outs of administrative protests and legal appeals. Biofuels firms lacking experience as military contractors may be less comfortable doing business with an agency that is large, labyrinthine, often opaque, and not very predictable.

Pennisi, "DARPA Offers $30 Million to Jump-Start Cellular Factories," *Science*, Vol. 333, July 8, 2011, p. 147.

7

US GOVERNMENT ALTERNATIVE FUEL POLICIES

Governments in many countries subsidize energy. Some sell cheap gasoline to mollify restive populations. Others underwrite nuclear power for reasons of national prestige. Government policies created the biofuels industry. In the absence of subsidies, there would be no significant production of fuel ethanol in the United States or Brazil, of biodiesel in the United States or Europe. Fear of rising oil prices and possible shortages, along with lobbying by agricultural interests, helped foster these policies, just as fear that the world would run out of oil underlay earlier policies supporting synthetic fuels made from coal, promoted in the United States by the Interior Department long before DOE existed. National security has been a common rationale.

The US government first extended tax preferences for oil exploration in 1916. Tax expenditures remain popular because they are less visible than other forms of subsidy, such as the quotas on imported oil that protected domestic firms from low-priced competition during the 1960s. Congress is unlikely to end energy subsidies.

Biofuels policies should be evaluated, and subsidies justified, on their own terms in a context of US national interests.

These interests have come to include the protection provided by US military forces to oil exporting countries and trade routes. The United States, while supplying aid to Egypt and Israel, otherwise had little presence in the Middle East until the 1979 Soviet invasion of Afhganistan triggered fears of further incursions by the Red Army into oil-rich regions of the Persian Gulf. President Jimmy Carter warned in January 1980 that "The Soviet Union is now attempting to consolidate a strategic position, therefore, that poses a grave threat to the free movement of Middle East oil."[1] Much later it became clear that Kremlin leaders had stumbled into Afghanistan without anything resembling strategic intent.[2] In any event, the Soviet Union vanished more than 20 years ago and no remotely comparable military power exists to threaten Gulf oil resources today, yet the

[1] Enunciating what became known as the Carter Doctrine in his January 23, 1980 State of the Union address, the president went on to say, "An attempt by any outside force to gain control of the Persian Gulf region will be regarded as an assault on the vital interests of the United States of America, and such an assault will be repelled by any means necessary, including military force." *Public Papers of the Presidents of the United States: Jimmy Carter, 1980-81: Book I – January 1 to May 23 1980* (Washington, DC: Government Printing Office, 1981), p. 197.

[2] "Soviet Invasion of Afghanistan Was Not a Grand Design But a Grand Entanglement Resulting from Faulty Intelligence, Excessive Secrecy, and a Paralyzed Leadership, According to Conference of Former Decision-Makers," National Security Archive Electronic Briefing Book No. 409, December 21, 2012, <www.gwu.edu/~nsarchiv/NSAEBB/NSAEBB409/>.

US military presence remains, a shield protecting supplies of oil flowing to allies including Germany and Japan.

Governments recognize that demand for energy is highly inelastic and that supply interruptions threaten large-scale disruption. Rising prices also stir popular protest, and in the United States political sniping. Otherwise higher prices could be seen as desirable to foster efficiency-enhancing innovations, discourage wasteful consumption (commonplace in countries where governments keep energy prices artificially low), and slow CO_2 emissions. Of course high prices also encourage exploration for new fossil fuel reserves. Meanwhile, at times of low prices, subsidies for conservation and renewables, including reserved markets such as that created by RFS2, draw criticism for interfering with market mechanisms, nowhere more than in the United States.

Policymakers generally realize that ordinary economic markets do a poor job of anticipating the always inscrutable future. At least implicitly, then, alternative fuels policies can be viewed as helping map transition pathways away from oil in anticipation of tightening supplies and rising prices. Innovation takes time. If oil prices do begin to disrupt economic activity, given a portfolio of alternatives in varying states of readiness market forces could begin to take over, stimulating further technology development and steering adaptation. Lacking advance preparation, it might take years of R&D and scale-up to bring needed production capacity into existence. In this view, governments, aware of the short-sighted nature of economic markets, intend to mark out and plan for a future in which market mechanisms facilitate final choices among fuels technologically proven and available for scale up with rising demand.

Over the past several decades, a second concern has emerged: global warming. While the effects of climate change cannot be predicted with confidence, they promise to be severe in some regions, and to increase the risks of international conflicts and thus the chances that the nation's military will be called upon to intervene somewhere in the world as so many times in the past. Because CO_2 released by combustion of fossil fuels is the predominant greenhouse gas, it is hard to envision a future in which replacement of fossil fuels does not become a principal means of managing climate change. The first question, then, concerns the extent of worldwide fossil fuel resources.

Energy Costs, Supplies, and Prices

Notwithstanding the runnup in oil prices since 2007, energy remains relatively inexpensive and fossil fuels abundant. Oil production in North America has increased and US imports have declined with exploitation of unconventional sources including oil sands, oil shales, and tight oil. Many of these sources were inaccessible or too costly to tap before innovations such as horizontal drilling and hydraulic fracturing, fracking for short. Much the same family of technical tools has increased supplies of natural gas.

Unlike the oil market, which is global, natural gas markets are regional, since gas moves cheaply only by pipeline. Even in the United States a good deal of natural gas released from oil-bearing formations, continues to be flared, or burned off at the well, in the absence of any good way to get it to customers.[3] With

[3] "Over One-Third of Natural Gas Produced in North Dakota Is Flared or Otherwise Not Marketed," Energy Information

the North American market largely isolated from other centers of demand, domestic prices have dropped precipitously. Natural gas has many uses, reserves will begin to decline at some point, and prices will rise. Yet these events appear to be well off. In his 2012 State of the Union address, President Barack Obama said, "We have a supply of natural gas that can last America nearly 100 years."[4] That seems no exaggeration (Box D).

Exploitation of unconventional petroleum liquids and natural gas suggests a world that is far from running out of oil or oil substitutes. Fossil fuels—coal, oil, natural gas—almost certainly will be extracted and burned for decades to come. Poorer countries especially will consume large tonnages of coal, the cheapest fossil fuel by far, and more of that coal will come from the United States.[5] Objections to fracking and to the environmental damage caused by extracting and processing oil shales, oil sands, and tar sands will no doubt continue, delaying some production. But opponents will have difficulty

Administration, November 23, 2011, <www.eia.gov/todayinenergy/detail.cfm?id=4030>.

[4] "Remarks by the President in State of the Union Address," White House, January 24, 2012, <www.whitehouse.gov/the-press-office/2012/01/24/remarks-president-state-union-address>.

[5] The International Energy Agency reports that coal accounted for more than half of worldwide increase in energy consumption during the past decade, "growing faster even than total renewables." *World Energy Outlook 2012: Executive Summary* (Paris: International Energy Agency, November 2012), p. 5. With declining domestic consumption, US coal exports more than doubled over the past half-dozen years. *Quarterly Coal Report, October-December 2012* (Washington, DC: Energy Information Administration, March 2013), Table 4, p. 10.

overcoming the political power of energy companies and energy lobbies, which gain support from the public's appetite for (and the economic benefits of) low-cost gasoline and electricity and also, in the United States, from landowners who lease mineral rights to energy companies.

The EIA's Reference case (business-as-usual) has crude oil increasing from around $100 per barrel currently to the vicinity of $160 per barrel in 2040; other analysts arrive at somewhat lower prices (and still others put forward alarmist scenarios).[6] While US consumption of petroleum has declined since the middle of the last decade, in part because of recession, demand in rapidly growing economies such as that of China will continue to prop up prices. Recognition that oil will continue to be available, even though prices may rise above $200 per barrel, perhaps well above, should dampen concerns that supplies of petroleum fuels will simply dry up.

For policymakers, a future of available fossil fuels—some of them cheap (coal) and even the most costly (oil) not exorbitantly expensive—aggravates the dilemmas of global warming. A true "end of oil," pushing prices steeply upward, would seed much innovation in substitutes; market forces would then drive the best of these through commercialization and into the ongoing

[6]*Annual Energy Outlook 2013,* DOE/EIA-0383(2013) (Washington, DC: Energy Information Administration, April, 2012), Fig. 49, p. 58. For forecasts by other organizations, see *Annual Energy Outlook 2012,* DOE/EIA-0383(2012) (Washington, DC: Energy Information Administration, June 2012), Table 13, pp. 104-105.

stages of continuous improvement.[7] That sort of world does not seem near. Repeatedly, warnings of coming oil shortages and price jumps have been followed by new discoveries and reclassifications that move known reserves from unrecoverable to recoverable categories.[8]

To the extent that new reserves continue to appear, pressures for migration away from petroleum will not build much, with adverse implications for both GHG emissions and the biofuels industry. More so than any other nation, *the United States – because it has so much natural gas and unconventional oil, and so much capacity to innovate – should ask how best to foster alternative energy in the face of weak economic forces.* Critics issue frequent reminders that Washington's energy policies lack consistency and coherence. That is probably inevitable

[7] Economies appear to have become more resilient in the face of fluctuating energy prices than in the 1970s. See, e.g., Lutz Kilian, "The Economic Effects of Energy Price Shocks," *Journal of Economic Literature*, Vol. 46, No. 4, 2008, pp. 871-909. That very resilience may tend to reduce the impetus for innovation. Recent work also suggests that rising energy prices may not by themselves induce the technological innovations needed to decarbonize the world economy. Daron Acemoglu, Phillipe Aghion, Leonardo Bursztyn, and David Hemous, "The Environment and Directed Technical Change," *American Economic Review*, Vol. 102, No. 1, 2012, pp. 131-166. The model developed by these authors, furthermore, shows that delay in stimulating innovation through non-price policies will increase long-run costs by extending the period of slowdown in economic growth associated with the eventual shift to technologies having dramatically reduced greenhouse impacts. Such results run counter to arguments by the many economists who contend that delay is the low-cost policy.

[8] James D. Hamilton, "Historical Oil Shocks," Working Paper 16790, National Bureau of Economic Research, Cambridge, MA, February 2011.

given the way governance works in the United States—and not just for energy but across much of the R&D and "industrial policy" spectrum.[9] Less forgivable has been the tendency to believe some sort of radical innovation could make a big difference. R&D outcomes can be predicted, broadly speaking; radical innovation cannot. Policymakers should not count on breakthrough discoveries to address pressing societal problems such as global warming. The better course is to seek innovations continually, big and small, from many sources.

Box D

Supply and Demand for Oil and Natural Gas

High oil prices and technologies such as fracking have spurred domestic production of fossil resources including tight oil—liquid-phase petroleum found in deep, low-permeability geological formations such as shale and chalk. (Sometimes called shale oil, tight oil is also found in non-shale formations.)

Between 2005 and 2010, the Energy Information Administration (EIA) raised its forecasts for technically recoverable tight oil by an order of magnitude, from 3.7 billion barrels to 33 billion barrels.[a] In EIA's most recent business-as-usual "Reference case," tight oil output rises to 2.8 million barrels per day in 2020 and shortly thereafter begins a slow decline to a level just above 2 million barrels

[9] Craufurd D. Goodwin, ed., *Energy Policy in Perspective: Today's Problems, Yesterday's Solutions* (Washington, DC: Brookings, 1981) includes many rigorously documented examples of the ways in which the politics of energy dominated energy policy from the Truman administration to that of Jimmy Carter. It would be hard to argue that much has changed since.

per day in 2040.[b] These are substantial volumes considering current US production of oil from all sources of around 6.5 million barrels per day.

Estimates and projections of US and North American reserves could increase further in coming years, perhaps substantially. Canada exports more oil to the United States than does any other country and has a great deal of oil sand and shale. (In contrast to tight oil liquids, oil shale designates shale and shale-like formations containing organic solids; heating liquefies the organics, which can then be extracted.) US and Canadian energy markets are substantially integrated and expanding fossil resources in both countries will exert at least some downward pressure on oil prices. This is not to say that prices will decline, simply to note one of several reasons why they may not rise steeply. Profitable exploitation of unconventional oil in the United States reportedly requires oil prices in the range of $50-$70 per barrel and prices will likely remain above that level.[c]

Natural gas reserves have increased much more dramatically than oil reserves, with estimates of technically recoverable domestic gas ranging above 4000 trillion cubic feet (tcf).[d] Annual imports (3.1 tcf) and exports (1.6 tcf) are small compared to production (23 tcf) and consumption (25.5 tcf).[e] Consumption is not expected to increase greatly. EIA expects production of natural gas to rise through 2040, the end point in the agency's latest projections, with exports rising to surpass imports before 2020.[f] For such reasons, President Obama's reference to "nearly 100 years" supply seems entirely plausible.

US natural gas prices plummeted with increases in supply. In the middle of the last decade domestic natural gas cost nearly as much on an energy-equivalent basis as crude oil; today with oil prices up and natural gas prices down, oil costs roughly five times more for equal energy content.[g]

Like oil, natural gas moves easily and cheaply through pipelines. Unlike oil, which also moves cheaply worldwide by sea, in the absence of pipelines gas must be transported under high pressure in specially designed tanks or at cryogenic temperatures after liquefaction. Even at the highest practical pressures, the volumetric energy density of compressed gas is low. Large-scale liquefied natural gas (LNG) facilities cost billions of dollars.[h] With high upfront costs for moving gas to points of consumption—whether for pipeline construction or LNG facilities—markets have been regional rather than global and large price differentials common. In some parts of the world natural gas now commands prices two or three times higher than the US price, making exports very attractive. Facilities built a few years ago to receive LNG from the Middle East are being "spun around" to serve as export terminals.[i] By the end of 2012, more than 20 companies had requests for export permits, required for shipping gas abroad, pending or recently approved in Washington.[j]

---

[a] *Annual Energy Outlook 2012,* DOE/EIA-0383(2012) (Washington, DC: Energy Information Administration, June 2012), Table 14, p. 57. Technically recoverable reserves assume current technology and disregard costs.

[b] *Annual Energy Outlook 2013,* DOE/EIA-0383(2013) (Washington, DC: Energy Information Administration, April 2012), Figure 97, p. 82. EIA's Reference case assumes a continuation of current policies and rates of technological advance.

[c] Ed Crooks, "Oil Hopes in 'Saudi America' are Unlikely to be Fulfilled," *Financial Times,* December 3, 2012, p. 12.

[d] *Prudent Development: Realizing the Potential of North America's Abundant Natural Gas and Oil Resources* (Washington, DC: National Petroleum Council, 2011), p. 69 and Table 1-3, p. 72. The Council, an advisory body to the Secretary of Energy, draws much of its membership from industry. Proved domestic reserves, about 350 tcf and defined as "estimated volumes expected to be produced, with reasonable certainty, under existing economic and operating conditions," do not come close to the council's figure and other high estimates. *US Crude Oil and Natural Gas Proved Reserves, 2011* (Washington, DC: Energy Information Administration, August 2013), Table 3, p. 12. Definition in *Annual Energy Outlook 2012*, p. 56.

[e] Few estimates for US natural gas consumption over the next several decades exceed 30 tcf per year, according to *Annual*

Energy Review 2011, DOE/EIA-0384(2011) (Washington, DC: Energy Information Administration, September 2012), Table 19, p. 62. For imports, exports, and production in 2012, see *Monthly Energy Review April 2013*, DOE/EIA-0035(2013/04), Table 4.1, p. 69. Most exports and imports move between the United States and Canada, which are linked by pipeline. Mexico also buys considerable US natural gas.

[f] *Annual Energy Outlook 2013*, Figure 89, p. 78.

[g] *Annual Energy Review 2011*, Table 5.18, p. 161 and Table 6.7, p. 191.

[h] Brian O'Keefe, "Exxon's Big Bet on Shale Gas," *Fortune*, April 30, 2012, pp. 75-85, reports figures of $16 billion to over twice this sum for the largest terminals.

[i] Richard Martin, "America's New Job Machine Is Heating Up," *Fortune*, April 30, 2012, pp. 89-92; quoted phrase on p. 92. "Special Report: Natural Gas, An Unconventional Bonanza," *Economist*, July 14, 2012, provides a useful review of international gas markets.

[j] "Applications Received by DOE/FE to Export Domestically Produced LNG from the Lower-48 States," <http://www.fossil.energy.gov/programs/gasregulation/index.html>. The Natural Gas Act of 1938, which views natural gas as a "natural monopoly" much as other utilities were treated at the time, regulates exports "consistent with the public interest." Michael Ratner, Paul W. Parformak, and Linda Luther, *US National Gas Exports: New Opportunities, Uncertain Outcomes*, R42074 (Washington, DC: Congressional Research Service, November 4, 2011), p. 13. Because gas explosions are so dangerous, transport and storage are also closely regulated.

Biofuels R&D

There are a great many ways to make biofuels, some proven, others demonstrated at laboratory scale, and still more simply hypothesized. Leaving aside corn ethanol and biodiesel, little information on costs is publicly available, or even exists, since production scale plants have yet to be built. More R&D and more operating experience will be needed to clarify choices of "production systems"—i.e., combining processing technology and feedstocks, including cultivation and collection methods. Given this relative immaturity, major tasks for R&D still concern *selection*: surveying candidate systems, screening them to identify the most promising, and deepening the knowledge base to support their development. Refinement and continuous innovation will come later.

In the usual division of labor, private firms undertake technology development with commercialization in view. Selection-focused R&D merits government support because much or most of the work will be too speculative to attract private funding. DOE's Biomass Program has supplied most of the federal government's research support, supplemented by USDA funds and more focused R&D supported by the Advanced Research Projects Agency-Energy (ARPA-E).

In recent years the Biomass Program has distributed in the range of $200 million in R&D funds annually; DOE's fiscal 2013 budget request seeks $270 million for the program.[10] DARPA's Biofuels Program included some work on jet fuel from algae. For DOE, algal feedstocks and processing have been a central focus; roughly half of Biomass Program funding during fiscal

[10] "Energy Innovation Tracker," <energyinnovation.us/?s= energy+department+biomass+program>. USDA has recently spent around $70 million annually on biofuels R&D. Keith O. Fuglie, et al., *Research Investments and Market Structure in the Food Processing, Agricultural Input, and Biofuel Industries Worldwide,* Economic Research Report Number 130 (Washington, DC: Department of Agriculture, Economic Research Service, December 2011), Table 10.3, p. 134.

2009-2011 supported such work.[11] This is not a new thrust. Starting in the late 1970s, DOE funded 15 years of R&D under its Aquatic Species Program in the course of which scientists collected and studied over 3000 algae strains in search of candidates for transportation fuels.[12] The program ended in the mid-1990s, a casualty of low oil prices. The renewal of interest and support reflects two potential sources of advantage for algae compared to other sources of biomass. Growing in water, algae need not encroach on arable land. Second, many strains grow very rapidly, with photosynthetic efficiencies—the fraction of energy in sunlight converted into chemical energy—several times that of typical agricultural crops.

Although a few firms have placed early bets, commercialization and large-scale production of algal fuels appears well in the future. The Aquatic Species Program only scratched the surface and much more work will probably be needed to uncover the best strains. Scientists have identified over 40,000 naturally occurring photosynthesizing forms of microalgae and

[11] *Biomass Program: 2011 Biennial Review Report* (Washington, DC: Department of Energy, Energy Efficiency & Renewable Energy, February 2012), p. 2. Also see *Biomass Multi-Year Program Plan* (Washington, DC: Department of Energy, Energy Efficiency & Renewable Energy, April 2012).

[12] *A Look Back at the U.S. Department of Energy's Aquatic Species Program: Biodiesel from Algae*, NREL/TP-580-24190 (Golden, CO: National Renewable Energy Laboratory, July 1998). For a brief and more recent review, see René H. Wijffels and Maria J. Barbosa, "An Outlook on Microalgal Biofuels," *Science*, Vol. 329, August 13, 2010, pp. 796-799. More comprehensive surveys include Al Darzins, Philip Pienkos, and Les Edye, *Current Status and Potential for Algal Biofuels Production: A Report to IEA Bioenergy Task 39,* Report T39-T2 (N.p.: International Energy Agency, August 6, 2010).

cyanobacteria, with more to be found.[13] Gene sequences have been determined for only a tiny fraction, perhaps a few hundred. Sifting candidates and seeking improvements through selective breeding and genetic engineering seems likely to be a major R&D task for some years to come. Investors will be understandably wary of picking a strain for commercialization when something better might turn up next month or next year.

Consumption of water and other inputs could be serious obstacles to large-scale production. Recently an expert committee of the National Research Council concluded that "scale-up of algal biofuel production sufficient to meet at least 5 percent of U.S. demand for transportation fuels would place unsustainable demands on energy, water, and nutrients with current technologies and knowledge."[14] The mention of "demands on energy" signals a major drawback of algae, given the current state of technology: the energy content of the fuel may be little more than energy inputs for growing and processing. Despite high photosynthetic efficiency, most assessments of algae on this basis—sometimes called energy return on investment—indicate meager positive margins, little if any better than corn ethanol.[15]

As the dauntingly open research agenda for algal

[13] *Sustainable Development of Algal Biofuels* (Washington, DC: National Academies Press, 2012), p. 27.

[14] *Sustainable Development of Algal Biofuels*, p. 2. The quotation continues: "However, the potential to shift this dynamic through improvements in biological and engineering variables exists."

[15] *Sustainable Development of Algal Biofuels*, pp. 126-129; for comparisons with non-algal biofuels, see Fig. 6-3, p. 201.

fuels suggests, there can be no assurances that private firms will find optimal or near-optimal packages of feedstock and processing technologies through their own self-financed efforts. *DOE and DoD, especially DARPA, should refocus on supporting the technology base for biofuels.* DOE has money for biofuels R&D. That money might be spent more effectively with the benefit of the mission-oriented discipline brought by DoD to planning, project selection, and oversight.[16] The hundreds of pages of description and commentary released in conjunction with DOE's most recent set of Biomass Program platform reviews illustrate both the strengths and the weaknesses of the agency's approach.[17] Strengths include the ability to identify and provide consistent funding for high quality science.[18] Weaknesses include the claims of DOE's own national laboratories on research dollars and the inbreeding that is hard to avoid when research groups and reviewers come from the same or closely related technical communities. ARPA-E, set up in part to break some of these patterns, and to help the rest of DOE revitalize itself, has also been supporting biofuels R&D, chiefly its PETRO and

[16] We summarize the benefits the military mission brings to R&D management in *Energy Innovation at the Department of Defense: Assessing the Opportunities* (Washington, DC: Consortium for Science, Policy and Outcomes, March 2012); see especially pp. 2-7 and 13-14.

[17] *2011 Platform Review Report: An Independent Evaluation of Platform Activities for FY 2010 and FY 2011* (Department of Energy: Biomass Program, February 2012).

[18] DOE does fund some work directed at demonstration, commercialization, and scale-up. As of mid-2012 the agency lists support for 24 biorefinery projects. *Integrated Biorefineries: Biofuels, Biopower, and Bioproducts*, DOE/EE-0767 (Washington, DC: Department of Energy, Biomass Program, July 2012).

Electofuels programs, each funded in the range of $50 million.[19] DOE could also learn from DoD's sensitivity to customers, even though these customers, the armed forces, are part of the public sector. DoD managers understand that new technologies must function reliably and effectively under harsh conditions in the hands of ordinary military personnel. They also understand how to work with industry, which carries out the bulk of the R&D funded by defense agencies.

Natural Gas-to-Liquids as a Transition to Biofuels

Transportation fuels and coal-fired electrical power plants produce two-thirds of the nation's energy-related CO_2 emissions.[20] These should be major targets of public

[19] See <arpa-e.energy.gov/?q=arpa-e-programs/petro> and <arpa-e.energy.gov/?q=arpa-e-programs/electrofuels>. Both these programs fund 10-12 projects, mostly at universities. PETRO, which stands for Plants Engineered to Replace Oil, supports work aimed at direct production of biofuels from plants, avoiding costly and energy-intensive processing steps such as gasification. The Electrofuels program seeks to cut out a different step in the biofuels production chain, that of photosynthesis, which is not very efficient, by means of microorganisms (e.g., bacteria) that can turn other non-solar forms of energy into transportation fuels. As for most ARPA-E programs, the intent is "to make revolutionary advances in breakthrough sciences, translate scientific discoveries and cutting-edge inventions into technological innovations, and accelerate transformational technological advances in areas that industry by itself cannot and will not support because of technical and financial risk and uncertainty." *Department of Energy FY 2012 Congressional Budget Request*. DOE/CF-0069, Volume 3 (Washington, DC: Department of Energy, Office of the Chief Financial Officer, February 2011), p. 434.

[20] Adjusted for global warming potential, CO_2 released in

policy because other sources are small and numerous by comparison. As technological and policy issues the two are very different. Each of 580 coal-burning power plants emits large volumes of CO_2.[21] Each of 250 million or so cars and trucks makes a near-negligible contribution to CO_2 emissions even though their collective impact is nearly as great.[22]

Utilities have been retiring their oldest, least efficient (hence least profitable), and most polluting (risking clean air violations) coal plants. With natural gas cheap and capital costs for building gas-fired plants far below those for coal-burning facilities, gas has provided most

burning fossil fuels makes up nearly 80 percent of US GHG emissions. Transportation (all modes) accounts for one-third (32.4 percent) of the CO_2 from fossil fuels, with something over three-fifths of this from personal vehicles. Electric power generation accounts for 42 percent of CO_2 from burning fossil fuels, four-fifths of this (80.9 percent) from coal and most of the remainder from natural gas (and coal therefore accounting for 34 percent of all fossil fuel-related CO_2). *Inventory of U.S. Greenhouse Gas Emissions and Sinks,* EPA 430-R-12-001 (Washington, DC: Environmental Protection Agency, April 15, 2012), Table ES-2, pp. ES-4 - ES-6, ES-8 and 3-13 (personal vehicles), and Table 3-5, p. 3-4 (electric power).

[21] These power plants house about 1400 steam generating units. *Electric Power Annual 2010* (Washington, DC: Energy Information Administration, November 2011), Table 1.2, p. 7 and Table 5.1, p. 49.

[22] "State Motor-Vehicle Registrations - 2011 (Revised)," Department of Transportation, Federal Highway Administration, Table MV-1, March 2013, <www.fhwa.dot.gov/policyinformation/statistics/2011/mv1.cfm>. The number of registered vehicles fell slightly during the post-2007 economic slump.

of the replacement capacity.[23] Policymakers, should they decide to accelerate the ongoing shift from coal to natural gas and renewables, have the tools to do so—and indeed have begun—and also to require that CO_2 from existing fossil fuel plants, including those fired by natural gas, be captured and sequestered to keep it from the atmosphere.[24] No similar remedies are available for motor vehicles. Trapping the CO_2 they emit is impractical; only by replacing existing cars and trucks with new vehicles or by changing the fuel can the forcing effects on climate be mitigated.

Federal policies have so far centered on CAFE (Corporate Average Fuel Economy) standards. Saving fuel reduces CO_2 emissions, but that was not the original intent. Congress passed the enabling legislation in 1975 as part of the response to the 1973-74 Arab oil embargo. Washington has also sought to foster powertrain innovations, most recently battery-electrics. These displace CO_2 emissions from vehicles to electrical utilities. With a growing fraction of the nation's electricity from low-carbon sources, and with increases in overall energy conversion efficiency (i.e., over the life cycle, from the original energy source to the highway), average CO_2 output per vehicle declines.

Neither CAFE nor powertrain innovations can make

[23] *Monthly Energy Review April 2013*, DOE/EIA-0035(2013/04) (Washington, DC: Energy Information Administration, April 25, 2013), Table 7.2a, p. 95. In the middle of the last decade, coal-burning plants produced about half of US electricity and natural gas-fired plants 19 percent. By 2012, coal's share had fallen to 37 percent and gas accounted for 30 percent.

[24] *Innovation Policy for Climate Change* (Washington, DC: Consortium for Science, Policy and Outcomes, September 2009), especially pp. 12-14.

much difference in the short to medium term. The nation's motor vehicle fleet is too big and turns over too slowly. The market has fragmented, even fissioned, with still growing numbers of nameplates and models targeting a multiplying set of niches. Product differentiation strategies coupled with declines in minimum efficient scale for design and manufacture mean that automakers expect profits on models that sell in the tens of thousands rather than, as formerly, the hundreds of thousands. Without draconian regulations, many of the choices popular among Americans, big pickup trucks and SUVs especially, seem unlikely candidates for battery-electric power no matter the technical advances of the next two or three decades. Government policies encouraged these shifts in customer choice by setting lower mileage (and safety) standards for light trucks, a regulatory (and political) rather than a functional classification that includes most vans and minivans, SUVs, and pickup trucks. So long as these met the light truck guidelines, automakers could design and build vehicles with the spacious interiors and high horsepower that many Americans have wanted while avoiding the costs of also providing high fuel economy. The US fleet now includes 126 million passenger cars; trucks (personal vehicles and commercial trucks of all types and weights) make up the remaining 118 million.[25] With much of the nation's geography ill-suited to electric vehicles (the sprawling Great Plains with widely spaced population centers; the Mountain West with its battery-draining upgrades; the Northern states with winter temperatures that degrade battery performance even as energy must be diverted for defrosting and interior heating), the implication is that battery-electric vehicles will be one more niche, too

[25] "State Motor-Vehicle Registrations - 2011 (Revised)."

small to have much effect on CO_2.

Even if government's approach were better aligned with consumer preferences, average vehicle age has doubled since the mid-1970s, from 5-6 years to about 11 years, which has slowed penetration of advanced technologies including high-efficiency powertrains and "lightweighted" vehicles.[26] The only way to reduce GHG emissions in the near term is to change the fuel. Alternative fuels with low GHG footprints would at the same time complement reductions in per-mile CO_2 emissions through advances in vehicle technology that will reach the market as manufacturers comply with the next CAFE increment, taking effect beginning in 2017.

Ethanol at 10 percent or 15 percent makes hardly any impact on CO_2. Neither does biodiesel blended at low percentages. As for military equipment, alternative fuels for road vehicles must be drop-in replacements, compatible with existing cars, trucks, and infrastructure. Diesel fuel and jet fuel can be made though hydroprocessing/hydrotreating or via Fischer-Tropsch methods, as Sasol does in South Africa (and now plans to do in the United States) and Shell in Malaysia and Qatar (and perhaps in Louisiana in the future). Natural gas or biomass (or coal) can also be made into gasoline, although the costs are presumably greater, else FT synthetic gasoline would presumably have been

[26] "Average Age of Vehicle Fleet Reaches Record High, According to Polk," January 17, 2012, <www.polk.com/company/news/average_age_of_vehicles_reaches_record_high_according_to_polk>; Don Pickrell and Paul Schimek, *Trends in Personal Motor Vehicle Ownership and Use: Evidence from the Nationwide Personal Transportation Survey* (Cambridge, MA: Department of Transportation, Volpe Center, April 23, 1998), Table 12, p. 18.

commercialized other than in South Africa, or via methanol as an intermediate product (these various possibilities were earlier described in Box C). Life-cycle GHG impacts of gas-to-liquids processing may currently be little if any better than those of petroleum, but improvement should be possible. Most of all, GTL technology is available and proven on a large scale.

Burning natural gas releases less CO_2 per unit of energy than other fossil fuels—about half as much as coal, with petroleum midway between the two.[27] This is not the whole story because more energy is lost or consumed in GTL processing than in refining crude oil. The CO_2 associated with this energy must be taken into account, increasing life-cycle GHG emissions. (Emissions associated with feedstock production, transportation, and delivery of fuel to the point of consumption should also be included.) The incremental energy-related GHG releases could be avoided by burning natural gas directly as a vehicle fuel. This is easy enough technically, either in new cars and trucks or through retrofits, but, like any alternative other than drop-in liquid hydrocarbons, is an unlikely prospect except for fleet-managed vehicles (e.g., commercial delivery and over-the-road trucks). Few private owners would accept the costs of conversion and the inconvenience of refueling. Nearly 160,000 retail outlets sell gasoline in the United States; not even 500 offer natural gas.[28] With the US

[27] *Instructions for Form EIA-1605: Voluntary Reporting of Greenhouse Gases* (Washington, DC: Energy Information Administration, October 15, 2007), Appendix H, p. 142. Natural gas itself, being mostly methane, does great damage if it reaches the atmosphere, since methane has 100-year greenhouse forcing impacts more than 20 times those of CO_2.

[28] According to "Service Station FAQs," American Petroleum Institute,<www.api.org/oil-and-natural-gas-overview/consu

personal vehicle market breaking up into ever more niches, alternative powertrain technologies will find places in some of these - hybrids (and perhaps diesel hybrids), battery-electrics (e.g. in urban markets), some natural gas, perhaps some fuel cells - but penetration promises to be slow. Under the most propitious of political and economic circumstances a meaningful transition to personal vehicles with alternative powertrains high in efficiency and low in GHG footprint would take decades.

Life-cycle GTL emissions depend on processing variables and the sources of the energy consumed as process inputs, which could be natural gas itself or purchased power from coal- or gas-fired plants, perhaps from nuclear or renewable sources. Process details—yields (how much of the chemical energy in the original feedstock remains in the finished product), reaction rates, temperatures and pressures (which determine, for instance, heat requirements, energy consumed by pumps and compressors, and so on)—are confidential, part of the recipe for profits at companies with viable GTL technologies such as Sasol and Shell. Because of this, little is known about the carbon footprints of operating GTL plants. Published estimates have been based on modeling studies with assumed parameters, typically reflecting circa 1990s technology. These results have not seemed encouraging, generally putting GHG

mer-information/service%20station%20faqs.aspx>, gasoline is sold at some 157,000 US locations. EIA reports about 1000 compressed natural gas refueling stations, the majority serving fleets and fewer than 500 open to the public. "Access to Alternative Transportation Fuel Stations Varies Across the Lower 48 States," April 30, 2012, <www.eia.gov/todayinenergy/detail.cfm?id=6050#tabs_AltTransportFuelStations-2>.

releases somewhat higher than for petroleum.[29] But they do not tell us what can be achieved with today's best technologies or might be achieved in a few years.[30]

Natural gas has many uses, reserves will begin to decline at some point, and prices will rise. These events are not imminent. One way or another, much of North America's natural gas will probably be burned. If policymakers wished, the federal government could slow the growth of exports and reserve adequate supplies of domestic natural gas for electric power generation and GTL fuels. Some domestic gas will likely be processed into liquid transportation fuels regardless of policy. Although Sasol might cancel its recently announced Louisiana plant and Shell decline to proceed in that state, a number of other firms have been evaluating similar investments.[31]

[29] See, e.g., M.Q. Wang, *GREET 1.5 – Transportation Fuel-Cycle Model, Volume 1: Methodology, Development, Use, and Results,* ANL/ESD-39, Vol. 1 (Argonne, IL: Argonne National Laboratory, 1999), pp. 44-61; John J. Marano and Jared P. Ciferno, *Life-Cycle Greenhouse-Gas Emissions Inventory For Fischer-Tropsch Fuels* (N.p.: Energy and Environmental Solutions, LLC for Department of Energy, June 2001); Paulina Jaramillo, W. Michael Griffin, and H. Scott Matthews, "Comparative Analysis of the Production Costs and Life-Cycle GHG Emissions of FT Liquid Fuels from Coal and Natural Gas," *Environmental Science & Technology,* Vol. 42, No. 20, 2008, pp. 7559-7565.

[30] Eric McFarland, "Unconventional Chemistry for Unconventional Natural Gas," *Science,* Vol. 338, October 19, 2012, pp. 340-342

[31] Andrew Herndon and Brian Swint, "Shale Glut Becomes $2 Diesel Using Gas-to-Liquids Plants," Bloomberg News, October 25, 2012, <www.bloomberg.com/news/2012-10-24/nazi-technology-turns-cheap-shale-gas-into-sub-2-diesel-

With Shell's Qatari plant finally operating, at least some analysts believe it will be profitable even at oil prices as low as $40 per barrel.[32] Sasol claims a still lower break-even oil price for its GTL plant in Qatar, $25 per barrel.[33] Because processing natural gas into synthesis gas is simple and not that costly, anything close to these figures would be impossible for biofuels firms to match with production processes that start with by gasifying biomass, a step that reportedly accounts for over than half of total production costs.[34] Unless natural gas prices skyrocket relative to biomass feedstocks, then, costs for biofuels seem certain to remain well above those for GTL fuels, the more so as collection and transportation costs limit biorefinery scale.

energy.html>.

32 Andrew Callus and Peg Mckey, "Shell Ramps Up Qatari Gas-to-Diesel Money Machine," Reuters, August 28, 2012, <www.reuters.com/article/2012/08/28/shell-gtl-idUSL6E8JP1F420120828>.

33 John M. Broder and Clifford Krauss, "A Big, and Risky, Energy Bet," *New York Times*, December 17, 2012. This article also includes comments by observers skeptical of such investments.

34 P.L. Spath and D.C. Dayton, *Preliminary Screening – Technical and Economic Assessment of Synthesis Gas to Fuels and Chemicals with Emphasis on the Potential for Biomass-Derived Syngas*, NREL/TP-510-34929 (Golden, CO: National Renewable Energy Laboratory, December 2003).

8

INNOVATION LESSONS FOR BIOFUELS

The world needs advanced, sustainable biofuels, if at all possible—those that do not add net CO_2 to Earth's atmosphere and do not compete with food crops—as part of the response to global warming. At some point in the more distant future, biofuels may also be needed to supplement petroleum. Innovation has been slow. There are a great many possible pathways, no one can know at this point which choices will turn out to be good ones, costs remain high and net greenhouse impacts uncertain. While government will be a crucial actor in the future of biofuels, policymakers cannot assume that research alone will generate sufficient innovation to somehow dissolve the knotty techno-economic issues discussed in earlier chapters. Market forces will be needed too. Yet markets are almost universally subject to small or large imperfections, tend to be short-sighted, and will remain indifferent to unpriced environmental impacts. Market mechanisms can contribute to selection of the best routes to advanced biofuels, but policymakers cannot rely on them blindly or unconditionally any more than they can rely on science and research.

Breakthroughs

"It's not often that truly game-changing technology comes along in the funeral services."[1]

Since the late 1970s, Schumpeterian depictions of innovation have been elaborated and embroidered almost endlessly, with a proliferation of descriptors infiltrating advertising, popular culture, and management-speak. Yet only a little more is known today about the sources of invention than during the first half of the last century, when Schumpeter and others such as S.C. Gilfillan wrote on innovation.[2] Discrete inventions (the telephone) and scientific discoveries (penicillin) sometimes trigger major bursts of technological and societal change. In other cases, continuous incremental change, proceeding almost

[1] From an industry newsletter quoted in Norma Love, "Ashes to Ashes, Dust to Dust ... Liquid to Liquid?" *Virginian-Pilot* [Norfolk], May 11, 2008, describing an alternative to burial or cremation that entails "dissolving bodies in lye and flushing the brownish, syrupy residue down the drain."

[2] Joseph A. Schumpeter, *The Theory of Economic Development* (Cambridge, MA: Harvard University Press, 1934 [translation of 1911 German edition]); S. C. Gilfillan, *The Sociology of Invention* (Cambridge, MA: MIT Press, 1970 [1935]). Schumpeter was careful to distinguish between invention and innovation: "to carry any improvement into effect is a task entirely different from the invention of it .... Although entrepreneurs of course *may* be inventors ... they are inventors not by nature of their function but by coincidence and vice versa"; pp. 88-89, italics in original.

invisibly, spawns what appears in retrospect to be radical innovation, as illustrated centuries ago by the ship design innovations about which Gilfillan wrote and more recently by the Internet.[3]

Quite commonly, far-reaching episodes with major impacts on patterns of consumption and business outcomes begin with "normal" engineering design and development. Examples include the microprocessor and Chrysler's original minivan. Neither was radical in a technological sense, yet both opened large new markets. When Intel commercialized the microprocessor, the idea had been "in the air" for some years; Intel neither invented the concept nor conducted new research.[4] Chrysler designed what was essentially a downsized window van, a category that had been on the market for 25 years, presenting it as a new type of family vehicle.

Among the hardware technologies that spawned the Internet, hard disk drives (HDDs) show performance gains as spectacular as those of microprocessors.[5] IBM

[3] On the Internet, see, e.g., *Innovation Policy for Climate Change* (Washington, DC: Consortium for Science, Policy and Outcomes, September 2009), pp. 18-19.

[4] William Aspray, "The Intel 4004 Microprocessor: What Constituted Invention?" *IEEE Annals of the History of Computing*, Vol. 19, No. 3, 1997, pp. 4-15. For further discussion of this and other cases, see John A. Alic, "Everyone An Innovator," Fred Block and Matthew R. Keller, eds., *State of Innovation: The U.S. Government's Role in Technology Development* (Boulder & London: Paradigm, 2010), pp. 236-260, on which the disk drive example following draws.

[5] HDDs provided the core illustration of successive disruptive episodes of technological change in Clayton M. Christensen's best-selling *The Innovator's Dilemma: When New Technologies Cause Great Firms to Fail* (Boston, MA: Harvard Business School Press, 1997). See pp. 3-59.

commercialized the HDD in the mid-1950s. The first units sold for around $50,000 and stored roughly 50 megabits (50 x 10^6 binary bits) on an assemblage of 50 rotating platters, each 2 feet in diameter, housed in an enclosure the size of a refrigerator. Today HDDs cost less than $10, sell in the hundreds of millions annually, and hold over 100 gigabits (10^{12} bits) on a disk no bigger than a quarter. In terms of storage density (bits per square inch of disk surface), performance has increased by 100 million times over five decades. The gains stem from a multitude of innovations, most of them incremental but some quite fundamental. The 2007 Nobel Prize in physics went to two scientists, Albert Fert and Peter Grünberg, for independent discoveries two decades earlier of a phenomenon known as giant magnetoresistance (GMR), deemed Nobel-worthy in part because of applications beginning in the 1990s to disk drives. Even so, GMR led to a relatively small "blip" in the long-term trend of HDD performance gains. The significance of GMR lies not so much in that blip as in enabling the long-term trend to continue. Still, if the phenomenon had not been discovered, or indeed did not exist in nature, engineers more than likely would have found some other means, if not so elegant or cost-effective, to keep the HDD trend going. And applications to HDDs did not result directly from the Nobel-winning work in the 1980s; these have been credited to R&D at IBM's San Jose, California laboratory in the 1990s. Much of the engineering design and development, finally, including that needed to manufacture low-cost, high-quality drives incorporating GMR technology, was carried out in East Asian countries where production has been centered for the past several decades. IBM itself sold its disk drive operations to Hitachi in 2004, in large part because of the slender profit margins in this intensely competitive business.

Depending on point of reference, then, the same family of innovations may seem either radical or incremental. GMR had radical impacts physically on disk drive storage media, as a fundamentally different magnetic structure at microscopic and submicroscopic scales. Yet over a span of nearly six decades GMR seems incremental—only a blip, one among many. This can stand for the normal course of technological innovation, from aircraft design to energy production.

More consistently sensible government policies would result from more widespread understanding of the ways in which even the most radical innovations have their impacts once embedded in larger economic and social systems. Individual innovations, no matter how dramatic they may seem from the standpoint of engineering and science, sometimes recede almost to the point of invisibility. Or as one observer put the matter in the context of another "technology" entirely: "What is striking is that linear programming is the confluence of influences instead of a single distinctive insight. I suspect that this is typical of radical innovations."[6]

Selection of Innovations

Ongoing R&D and accumulating production experience will reveal wide variation in production costs and sustainability for biofuels. Selection will be based almost entirely on costs, and these costs cannot be known until large-scale plants have been designed, built, and operated. Market mechanisms are the best way to make the ultimate choices, but even if they were to

[6] Robert Dorfman, "The Discovery of Linear Programming," *Annals of the History of Computing*, Vol. 6, No. 3, July 1984, pp. 283-295; quotation from p. 294. Dorfman himself was one of the "discoverers."

incorporate environmental costs markets do not function well in the absence of reliable information.

Analogies between evolution in biological systems and technological systems, loose though they may be, can help structure the biofuels policy agenda. The basis for such analogies is the distinction between generation of variety and selection for survival. In living systems, random mutations generate variety. Some mutations increase the fitness of the organism or system of organisms, the ability to survive and to reproduce in a given environment. The analogy is imperfect because the variations that lead to technological innovation are not random but purposeful and because knowledge accumulates in both codified and implicit forms for transmittal to future generations, whereas the genome is the transmittal mechanism in biological systems. Students of technological change have found such analogies useful despite these differences.[7]

Supply and demand provide driving forces for generation of technological variety and economic forces then select survivors from among these variants. When businesses conduct R&D, conceptualize and design new goods or services, and develop the needed production processes, they contribute to technological variety. Selection takes place when firms offer these new products for sale in competition with other products that may or may not be close substitutes. Both biodiesel and renewable diesel, for example, substitute for petroleum

[7] George Basalla, *The Evolution of Technology* (Cambridge, UK: Cambridge University Press, 1988); Richard R. Nelson and Sidney G. Winter, "Evolutionary Theorizing in Economics," *Journal of Economic Perspectives*, Vol. 16, No. 2, Spring 2002, pp. 23-46.

diesel, but the attributes of biodiesel differ from those of petroleum diesel in ways meaningful to cusomters. Indeed, DoD bars use of biodiesel because of the risks of damage to military equipment. Renewable diesel, on the other hand, considered strictly as a fuel for powerplants, has attributes substantially identical to petroleum diesel. These two alternatives differ between themselves, and in comparison with petroleum diesel, in environmental impacts, although that has no direct consequences for purchasers and, absent government policy, would not affect choice.

Demand also generates variety. Independent companies such as Schlumberger supplied a steady stream of technical advances in oil exploration to paying customers; sensing demand, Schlumberger innovated to fill it. Drillers innovated too. Key steps in the development and application of fracking, hydraulic fracturing combined with horizontal drilling, can be apportioned to both government programs and to a private firm, Mitchell Energy.[8] Unlike Schlumberger, Mitchell was not looking for a package of technology and services to sell; it was looking for oil.

Screening and selection or winnowing of innovations, which leads to reduction in variety as some products and businesses fail while others succeed, sometimes proceeds in straightforward fashion. Purchasers choose new products, such as Chrysler's minivans, and other firms follow with similar offerings. Yet even this example of market-guided evolution was not that simple. Government policies affected the selection criteria because minivans had only to meet fuel economy and safety standards for light trucks, not the more

[8] "Special Report: Natural Gas, An Unconventional Bonanza," *Economist*, July 14, 2012, pp. 5-6.

demanding standards for passenger cars. Automakers saved money in compliance that could be put toward product upgrades, higher profits, or lower prices.

Government policies also deselect innovations. In the 1920s, Thomas Midgley and others associated with General Motors (GM) discovered the anti-detonation properties of tetraethyl lead and related compounds as gasoline additives. Amounting to rapid burning, spontaneous, premature, and uncontrolled, detonation limits both the power output and efficiency of gasoline-fueled engines. Lead compounds slow the precombustion reactions that cause detonation, an effect that shows up as a higher octane rating (determined by testing the fuel in an actual engine), a measure of resistance to detonation. With higher octane fuels, the compression ratios of engines could be raised, increasing horsepower and fuel economy. Military planes could fly higher, faster, and farther; road vehicles also gained in performance attributes valued by users. GM, not in the chemical business, patented the discovery and took part ownership in a new firm, Ethyl Corporation, set up to manufacture lead compounds and market them to petroleum refiners.[9] Highly toxic lead accumulated in the environment. After having been "selected in" by market forces lead compounds in gasoline were later "selected out" by regulatory policies, although not until overwhelming evidence of harmful effects overcame years of industry opposition.

Policy Implications

Worldwide, "an estimated 300 companies [are] engaged in second-generation biofuels research,

9 Alfred P. Sloan, Jr., *My Years with General Motors* (New York, NY: MacFadden, 1965), pp. 221-226.

working with dozens of feedstocks and processes."[10] Most of these enterprises will probably disappear. Some may simply fail; others will become acquisition targets or merger partners. A few may survive as independent entities. The overall dynamic, labeled creative destruction by Schumpeter and his followers, might just as well be termed "creation and destruction."

More so than large firms, small companies with limited internal research capabilities depend on technical knowledge generated through government funding. Firms pull technical knowledge originating in publicly-funded R&D into the organization, using it to generate technological variety. The government contribution is indirect but vital. By contrast, policies that draw alternative fuels into the marketplace speed *selection,* the winnowing of variety. Well-adapted innovations emerge and then advance through ongoing incremental improvement. For petroleum products these processes of technical and economic selection have been ongoing for more than a century. For synfuels produced from coal, they have proceeded almost as long but at a much slower pace. For second-generation cellulosic ethanol, haltingly sluggish development suggests that evolutionary processes have barely taken hold. The federal government's reliance on market pull through RFS2 would have benefited from sustained earlier support for generation of technical knowledge and technological variety through research. Seemingly unaware of how long innovation takes, for some technologies a matter of decades rather than years, policymakers cut public R&D funding for biofuels in the 1990s simply because oil prices had for the time fallen.

[10] Guy Chazan, "US Policy Aids Second Generation Biofuel Push," *Financial Times*, June 4, 2012, Special Report: Modern Energy, p. 3.

9

LOOKING AHEAD: ENERGY AND CLIMATE

A massive and highly interdependent system of systems, the energy sector will only be restructured by multiple innovations over lengthy time periods, pushed by R&D and pulled by market forces. Most of the innovations will be extensions of existing technologies. The questions for policymakers, keeping in mind the fundamental goal of reducing greenhouse gas emissions, become: Which technologies are likely to advance rapidly driven by market forces alone? Which technologies may not? How can market forces be reinforced absent strong price signals? How can R&D be reinvigorated and more effectively managed, again absent strong price signals? And how might DoD, a major customer for jet fuel with extensive R&D capabilities and a prominent contributor to past episodes of innovation in technologies including aviation, contribute?

The world is a long way from running out of fossil

fuels, in part because of advances in technical means for exploitation of deep ocean reservoirs and unconventional oil and gas, just the latest episode in a century-long history of technological innovation underlying expansion of known and exploitable hydrocarbon reserves. Energy markets have performed reasonably well for over a century, with a few hiccups. Since the most recent of those, in the mid and late 1970s, global energy markets have become more resilient. There seems no compelling reason to believe that these markets will not, for several decades to come, provide the United States and much of the rest of the world with supplies of petroleum adequate to meet demand at prices that purchasers will tolerate, at least in wealthier countries. Always in the past, fears of lasting energy shortages have proved to be overblown and now rising oil prices have stimulated production of unconventional oil and gas.

What energy markets cannot do is mitigate global warming. Indeed, the more effectively energy markets function, by tapping new fossil fuel sources in response to rising prices, the less the impetus for innovation in alternatives. Fossil fuels will continue to be extracted and burned, in the United States and globally. Exploration, extraction and refining will continue to be highly profitable, and this fact, combined with the fossil industry's sheer magnitude and technological sophistication, will continue to give it advantages in turning innovation into economic advantage. Demand will grow, in developing countries especially. Alternative energy technologies will continue to advance, costs will come down, and some fossil fuels will be displaced—or perhaps just supplemented—but GHG emissions will continue to rise. *Resilient energy markets promise to continue supplying huge volumes of carbon-intensive fuels, while at the same time the world needs huge reductions in* CO_2 *emissions to slow climate change: the conjunction makes decisive policy intervention essential.*

No matter how strong the policy push, restructuring will take time. Many companies are working to commercialize biofuels. The pacing events take place inside companies, in their R&D facilities, pilot/demonstration plants, and eventually in production-scale biorefineries. While government can push technological advance through research and pull through market incentives, policymakers must be aware of the dangers of premature closure on what could turn out to be second-best choices. Government policy should adapt itself to the reality that marketplace selection will be a messy, drawn-out affair. With much technological uncertainty and hardly any reliable information on production costs, almost regardless of technical advances, scale-up promises to be piecemeal and slow. Some firms will commercialize innovations successfully; others will try and fail. Policymakers resolved to take global warming seriously cannot wait for the biofuels industry to grow as it has been. Every day, the US transportation fleet burns some 550 million gallons of liquid fuels. Because it will take massive production of carbon-neutral or carbon-negative energy sources to make much of an impact on CO_2 emissions, and because expecting to move directly from petroleum to biofuels risks pushing the eventual shift repeatedly off into the future, *policymakers should consider mapping out a transition pathway to sustainable biofuels, those with zero or negative net GHG emissions.*

The best way to start is to bring on natural gas-to-liquid fuels in volume and then replace natural gas as a feedstock with biomass, over time and to the extent that biofuels demonstrate sustainability. *The transition should begin with diesel and jet fuel made from domestic natural gas, with regulations that impose transparent life-cycle GHG accounting on GTL plants and require initial parity with petroleum.* Staged reductions in GHG emissions would follow. At least some technical advances that reduce GHG output will also contribute to cost reductions: commonality of interest should help defuse opposition. This policy would take advantage of the vast technological capacities of companies in the extended petrochemical industry, including specialized suppliers and consulting firms. GTL production fits with the experience base and business strategies of many of these companies. With GTL as a complement to existing lines of business rather than a threat, and with profits in view, petrochemical firms would find more reasons to innovate and fewer reasons to obstruct.

Although DoD purchases much more fuel than the rest of government, in the world market it is simply one customer among many. The recent initiative under which the Navy, DOE, and USDA would contribute up to half the costs of investments in design and construction of biofuels production capacity, with Navy and DOE participation under the Defense Production Act, may not be a sufficient incentive to attract potential suppliers struggling with recalcitrant technologies, even if Congress appropriates full funding of $510 million. After all, the 2007 Energy Independence and Security Act created seemingly powerful inducements for production of cellulosic ethanol and other advanced biofuels without, as yet, leading to significant production capacity. While a rapid shift to advanced biofuels is unlikely no matter what DoD does, to help smooth the transition from petroleum to alternatives *the Pentagon should make it plain to industry that it will purchase drop-in fuels made from natural gas alongside biofuels*, provided life-cycle greenhouse gas emissions are no worse than those of petroleum (as already required by law for federal procurements). Commitment to purchase jet fuel made from natural gas would strengthen incentives for companies to build GTL plants in the United States.

The argument for natural gas as a transition is not really that biofuels technology is not "ready." That would be an oversimplification. The argument is that premature attempts at commercialization contribute relatively little to subsequent evolutionary technical advances and industry growth because selection pressures acting on the multitude of possible technical pathways too easily misfire, as they arguably did with corn ethanol and biodiesel. Accordingly, while viewing drop-in GTL transportation fuels as a transitional step, the federal government should support biofuels R&D at higher levels than in recent years, with funding shared by DOE and defense agencies. As a major customer for fuels – a prototypical "smart customer" by virtue of internal capacities that cover the full range of technical activities, not just research – DoD should have a strong managerial role. Sorting through the many technical prospects and pathways that remain unexplored should be the initial goal. DOE funds substantial research though its Biomass Program. But as illustrated by the agency's ineffectual efforts to foster motor vehicle innovation going back decades, DOE, fundamentally a science (and nuclear weapons) agency, struggles to bring understanding of consumer markets and behavior to its R&D policies. For DoD, by contrast, innovation is a means to an end. DoD managers understand users, even if the services as "customers" for new technologies do not behave like ordinary business or household purchasers. Defense agencies also understand how science finds its way into practically useful systems and equipment, including the role of developmental testing and user feedback in helping steer research and development. DoD has fostered major innovations in aerospace propulsion and in many technologies unrelated to fuels. Defense agencies could do so again with biofuels.

Appendix

FUEL TYPES AND PROPERTIES

A1. Hydrocarbon Fuels

Crude oils from Texas differ from those found in Saudi Arabia or Venezuela. Oil comes in thousands of varieties, each a mixture of thousands of hydrocarbon compounds along with nonhydrocarbon impurities.[1] During refining, the constituents in crude oil boil off or vaporize at temperatures that depend on molecular weight and structure. Pure substances such as water undergo changes in phase—from liquid to vapor, from liquid to solid—at a single temperature. Mixtures, hydrocarbons or not, pass through phase changes over a range of temperatures, allowing separation of petroleum (fractionation) into mixtures that have a narrower distribution of molecular species than the original crude. These fractions still include many hundreds of different compounds. Further refining steps, such as catalytic cracking, break big molecules into smaller ones on the

[1] For a straightforward introduction, see *Aviation Fuels Technical Review*, FTR-3 (Houston: Chevron Global Aviation, 2006).

way to final products such as gasoline (relatively light), heavier fuels, asphalt, and other chemical products. One of these heavier fractions, kerosene—a broad range of "distillates" originally designating hydrocarbons suitable for household furnaces, stoves, and lamps—encompasses both jet fuel and diesel fuel.

The distribution of molecular weights (and the range over which the fuel boils) characterizes the familiar fuel categories. Most of the molecules in gasoline have 4 to 10 carbon atoms, averaging to something like C_8H_{17}. Jet fuel consists largely of molecules with 10 to 15 Cs, averaging around $C_{11}H_{21}$. Diesel fuel is heavier still, around $C_{16}H_{32}$. Batches of fuel meeting the same specification—JP-8, for instance, one of several grades of jet fuel standardized by the US military (JP stands for jet propellant)—will differ chemically depending on the composition of the original crude and refinery processing. Synthetic jet fuel can be made from coal, natural gas, or biomass. While meeting all or most of the specifications for JP-8, it will differ systematically in chemistry from petroleum JP-8.

Most of the hydrocarbons in petroleum fuels fall into one of four main classes: paraffins, olefins, napthenes, and aromatics. The four classes are distinguished by the presence or absence of multiple carbon-carbon bonds—two carbon atoms may be joined by single, double, or triple bonds—and also by molecular structure. Paraffins have single-bonded linear or branched chain structures; lacking multiple bonds, they are known as saturated molecules. Olefins are structurally similar but unsaturated, with one or more doubly-bonded carbons. In napthenes and aromatics some or all of the carbon atoms form rings; the difference is that in napthenes all the carbon bonds are single while aromatics have some double (or aromatic) bonds.

Standard refinery operations such as hydroprocessing or hydrotreating pass free hydrogen through the liquid petroleum under pressure and in the presence of catalysts, resulting in chemical reactions that shift the proportions of the types of hydrocarbons present and remove impurities such as sulfur and nitrogen compounds (sulfur, for example, will combine with free hydrogen to form hydrogen sulfide, which can then be separated out). Hydroprocessing/hydrotreating also reduces olefin and aromatic content; with added hydrogen, carbon-hydrogen bonds replace multiple carbon-carbon bonds. This is desirable since olefins can pick up oxygen in storage, degrading fuel quality, and high aromatic content contributes to incomplete combustion, leading to air pollution and in extreme cases exhaust smoke. Some aromatic content is nonetheless desirable, as discussed below. The remaining appendix sections deal mostly with jet fuels and their effects on system design and performance.

A2. Fuel Properties and Specifications

In 1951 the Air Force settled on JP-4, a "wide cut" blend of gasoline and kerosene, so called because of specifications written to permit a considerable range of proportions, as its standard jet fuel. Fuel suppliers generally delivered JP-4 that was more than 50 percent gasoline and, like aviation gasoline itself highly flammable. The Navy, well aware after World War II of the dangers of shipboard fires, selected straight kerosene instead, issuing its specification for JP-5 the next year, 1952.

During the Vietnam War the two services had contrasting experiences with their chosen fuels. Air Force planes carrying JP-4 suffered far more damage from fires than naval aircraft fueled with less flammable JP-5. The Air Force began exploring alternatives to JP-4—adopting the Navy's fuel was out of the question—

and in 1979 issued the JP-8 specification. JP-8 is identical to commercial Jet A-1 except for additives such as corrosion inhibitors and stabilizers. All jet fuels in widespread use today—JP-5 and JP-8, commercial Jet A (a US standard) and Jet A-1 (the standard most other countries have adopted)—are in fact quite similar. As Table A1 shows, the Navy's JP-5 specification requires a "flash point"—an indicator of flammability—that is 40 F higher than the others, and freeze point temperatures also vary; otherwise the differences are minimal.

Standards documents prescribe how specified properties are to be measured, in some cases with methods originating in the 1920s and 1930s.[2] Both flash point and freeze point temperatures as defined in the notes to Table A1, for example, have meaning only in terms of the prescribed tests. Like the octane number of gasoline, which is determined by running a specially designed engine on the fuel being tested, flash point and freeze point are "operationally defined": they cannot be reduced to fundamental physics or chemistry and have significance only in terms of the test procedure.

Except as specifications restrict harmful impurities, they rarely cover chemistry. This would be impractical, given the wide-ranging composition of crude oils. Physical properties predominate in specifications—e.g., density (mass or weight per unit of volume) and energy density (energy content per unit of mass). Requirements may be stated as maximums, minimums, or an allowable range. Freeze point is given as a maximum temperature since the intent is ensure that fuel will not begin to solidify in the extreme cold of stratospheric flight.

2 *Aviation Fuels Technical Review*; *Handbook of Aviation Fuel Properties*, Third Edition, CRC Report No. 635 (Alpharetta, GA: Coordinating Research Council, 2004).

Airlines and ASTM International, the private body that controls consensus standards for commercial fuels in the United States, consider Jet A adequate for domestic flights. On long international flights, over arctic regions especially, planes fill their tanks with lower freezing Jet A-1 to minimize the danger of fuel turning to slush. The aviation authorities in many countries have chosen to adopt the standards of the United Kingdom's Ministry of Defence, the de facto guardian of the Jet A-1 specification.

Table A1

Selected Jet Fuel Specifications [a]

	Density (lb/gal) [b]	Hydrogen Content (minimum) (mass percent)	Final Boiling Temperature (maximum) (F) [c]	Gravimetric Energy Density (minimum) (Btu/lb)	Freezing Temperature (maximum) (F) [d]	Flash Point Temperature (minimum) (F) [e]
Aviation gasoline [f]	Not specified	Not specified	338	18,700	-72	Not specified
JP-4 [g]	6.27-6.7	13.5	518	18,400	- 72	Not specified
JP-5	6.58-7.1	13.4	572	18,320	- 51	140
JP-8	6.47-7.0	13.4	572	18,400	- 53	100
Jet A, Jet A-1	6.47-7.0	Not specified	570	18,400	Jet A: - 40 Jet A-1: - 53	100

Table A1 continued

[a] Aviation gasoline included for comparison. Standards cover many more specifications than those listed and generally give property values in metric units; these have been converted to English units and also rounded for consistency and convenience.

[b] At 60 F.

[c] The temperature at which the highest boiling constituents transform from liquid to vapor.

[d] The temperature at which the last wax crystals in previously frozen (solid) fuel transform to the liquid phase.

[e] The lowest temperature at which fuel vapors ignite in the presence of an ignition source.

[f] 100LL, commercial low lead.

[g] JP-4 has been largely superseded by JP-8 but remains in use.

Sources: *World Jet Fuel Specifications with Avgas Supplement*, 2008 Edition (Brussels: ExxonMobile Aviation, 2008); *Detail Specification: Turbine Fuel Aviation, Grades JP-4 and JP-5*, MIL-DTL-5624U NOTICE 1 (Patuxent River, MD: Naval Air Systems Command Fuels & Lubricants Division, November 4, 2008); *Detail Specification: Turbine Fuel, Aviation Kerosene Type, JP-8 (NATO F-34), NATO F-35, and JP-8+100 (NATA F-37)*, MIL-DTL-83133H w/ AMENDMENT 1 (Wright-Patterson Air Force Base, OH: Air Force Petroleum Agency, September 14, 2012); *Handbook of Aviation Fuel Properties*, Third Edition, CRC Report No. 635 (Alpharetta, GA: Coordinating Research Council, 2004).

The fundamental purpose of specifications is to match fuels with intended usage. Road gasoline must not foul the catalytic converters of cars and trucks; this is irrelevant for aviation gasoline, since aircraft engines do not carry converters. Diesel fuel must have sufficient "lubricity" to protect tight-fitting moving parts in fuel pumps and injectors. Because standards and specifications cannot be written to anticipate all contingencies, fuels must also demonstrate "fitness-for-purpose," typically through tests and trials with actual equipment. By running engines on test stands and then conducting flight tests, regulatory bodies seek assurances concerning "fuel characteristics which are needed for safe operation but are not directly evaluated via a specification test."[3] This sort of testing accounts for much of the time and money spent in certifying alternative fuels such as synthetic substitutes for JP-8 or Jet A-1. When DoD interest in alternative fuels picked up a few years ago, impelled by surging oil prices, the Air Force began its certification tests with the B-52 bomber. Each of these planes carries eight TF-33 turbofans, a powerplant designed by Pratt & Whitney in the 1950s. The Air Force Research Laboratory first conducted a series of rig and engine tests on 50:50 blends of synthetic and petroleum kerosene, sometimes known as semisynthetic fuel. Flight tests followed, at first with only two engines running on semisynthetic jet fuel and the other six fed regular JP-8.[4] Cold weather,

[3] Richard Striebich, et al., *Dependence of Fuel Properties During Blending of Iso-Paraffinic Kerosene and Petroleum-Derived Jet Fuel*, AFRL-RZ-WP-TR-2009-2034 (Wright-Patterson Air Force Base, OH: Air Force Research Laboratory, November 2008), p. 3.

[4] Laura L. Lundin, "Laboratory Leads the Way," *U.S. Department of Defense Transformation News*, November 2006, <www.defense.gov/transformation/articles/2006-

toxicity, and fire suppression tests, among many others, were a further parts of the agenda.

A3. Engines and Aromatics

Initial Air Force certification for the B-52 and other aircraft types restricted synthetics to 50:50 blends, as had earlier approvals granted the South African firm Sasol for its semisynthetic jet fuel. Blends containing at least 50 percent petroleum provide sufficient aromatic content to protect seals, gaskets, and other components in contact with fuel. Aromatic hydrocarbons—ring molecules with double-bonded carbon atoms—cause elastomeric (rubber-like) materials to swell, helping maintain seal integrity in engines, fuel systems, and ground equipment. In their absence, gaskets and seals can shrink and begin to leak, especially older components that may have hardened and perhaps begun to crack or otherwise deteriorate.[5] The Air Force encountered a great deal of trouble with fuel leaks during the 1980s and 1990s simply in shifting from high aromatic JP-4 to lower aromatic JP-8.[6]

Found in crude oil and reduced but not eliminated during refining, aromatics occur in substantial proportion in petroleum fuels. JP-8, for example,

11/ta111006b.html>.

[5] Pat Muzzell, Leo Stavinoha, and Rebecca Chapin, *Synthetic Fischer-Tropsch (FT) JP-8/JP-5 Aviation Turbine Fuel Elastomer Compatibility* (Warren, MI: US Army Tank-Automotive Research, Development and Engineering Center, February 2005).

[6] Charles R. Martel, *Military Jet Fuels, 1944 - 1987*, AFWAL-TR-87-2062 (Wright-Patterson Air Force Base, OH: Aero Propulsion Laboratory, November 1987), pp. 17-18.

averages around 18 percent aromatics.[7] Synthetic kerosenes, regardless of feedstock and processing, have essentially zero aromatic content (under 0.5 percent, usually well under).[8] It makes no difference whether the kerosene is made by Fischer-Tropsch (FT) methods from natural gas, gasified coal or biomass, or by hydroprocessing/hydrotreating of biomass without prior gasification to produce fuels known as HEFA, for hydroprocessed esters and fatty acids, or HRJ, for hydroprocessed renewable jet fuel. Regardless of the original feedstock, chemical reactions during processing break apart multiple carbon-carbon bonds. Singly-bonded atomic pairs (carbon-carbon and carbon-hydrogen) take their place. The end result is a fuel composed entirely of chain-like paraffins, often called SPK, for synthetic paraffinic kerosene. To protect equipment, SPK can be mixed with petroleum (e.g., 50:50) or treated with special additives.

While low aromatic content can mean leaky, potentially dangerous fuel systems, requiring extra maintenance, high aromatic content has its own drawbacks. Depending on the engine design, an excess of aromatics can cause exhaust smoke, not only a pollutant but a signal that can alert an enemy to the presence of an otherwise invisible plane; F-4s flown by the US military over Vietnam were notorious for their

[7] *Department of Defense Handbook: Aerospace Fuels Certification*, MIL-HDBK-510-1A(USAF) (N.p.: US Air Force, February 8, 2010), Fig. C-4, p. 74.

[8] Clifford A. Moses, *Comparative Evaluation of Semi-Synthetic Jet Fuels - Addendum: Further Analysis of Hydrocarbons and Trace Materials To Support Dxxxx* (Alpharetta, GA: Coordinating Research Council, April 2009). Processes can be tuned to produce other types of hydrocarbons, but SPK has been the preferred output.

smoke trails (a trait of their powerplants that fuel characteristics did nothing to reduce). Highly aromatic fuels also increase the likelihood of internal engine deposits likewise resulting from incomplete combustion. For such reasons, specifications for petroleum jet fuel typically set maximums (but not minimums) for aromatic content.

In gas turbines and jet engines, fuel burns continuously (after startup), sprayed from nozzles inside canister-like combustors.[9] The chemical reaction sequences involved may, depending on the fuel, proceed along 100 or more distinct pathways. Air and hot gases move downstream as these reactions continue, releasing heat. These flows must be carefully managed. Some of the air leaving the (upstream) compressor, for instance, bypasses the combustor, flowing between the outer surface of a concentric perforated liner and the combustor canister itself. The bypass air cools the liner and protects other components from the very high temperatures in the flame itself, which no metal can withstand. Slots in the liner also feed in air downstream to dilute the hot flowing gases. Heat transfer to this excess air and the liner walls reduces the average temperature at exit from the combustor and entrance to turbine stage, keeping the turbine vanes and blades from overheating. High peak temperatures are desirable for best performance; the higher the temperatures at turbine entrance the more power the engine will produce and the more efficiently it will convert the chemical energy in the fuel into useful power or thrust. But even the specially formulated superalloys from which combustors, their inner liners, and other high-

[9] See, in general, Arthur H. Lefebvre, *Gas Turbine Combustion*, Second Edition (Philadelphia: Taylor & Francis, 1999).

temperature components are made (and coated with exotic protective and insulating layers) cannot withstand the very high temperatures, around 3700 F, of the flame itself. This is why designers take such elaborate measures to manage flows and temperatures within the combustor.

If all combustion reactions go to completion, the exhaust will consist only of carbon dioxide and water vapor. If they do not, unburned hydrocarbons and intermediate reaction products, generally including carbon monoxide and particulate matter, pass through the turbine and into the atmosphere. Paraffins and napthenes burn rapidly and therefore cleanly. Aromatics and other hydrocarbons with double-bonded carbons burn more slowly (the double bonds must first be broken), contributing disproportionately to incomplete combustion, to internal engine deposits left by residual intermediate reaction products, and to soot and other particulates sometimes visible as exhaust smoke.[10] Excessive deposit buildup can harm engine performance and may require periodic teardowns for cleaning. Particulates such as soot—wholly or partially carbon or intermediate products of combustion—are health hazards in jet engine exhaust just as in diesel exhaust. While paraffins burn rapidly and SPK can be considered a clean fuel, highly aromatic fuels can equally be called dirty fuels. That is why specifications for petroleum jet fuels limit aromatic content, usually to 25 percent (by volume).

Combustion phenomena depend not just on the fuel

[10] See, e.g., R.A. Rudey and J.S. Grobman, *Characteristics and Combustion of Future Hydrocarbon Fuels*, NASA TM-78865 (Cleveland, OH: National Aeronautics and Space Administration, Lewis Research Center, 1978).

but on the design of the engine and its components (e.g., injection nozzles, combustor geometry, slots and baffles for mixing air and fuel) and also on operating conditions. These effects have been extensively studied, although some remain imperfectly understood. Engine manufacturers rely heavily on experience, proprietary technical data, and internally developed design guidelines (also proprietary). At their design points, which for an aircraft powerplant will normally be high-speed, high-altitude cruise, combustors are 99-plus percent efficient, meaning nearly all the fuel burns completely, maximizing power output and overall efficiency (incomplete burning wastes some energy, which exits the engine still in the form of chemical energy rather than heat). Under other operating conditions, and especially at low loads (descent, taxiing), combustor/combustion efficiency may be much lower and fuel properties tend to have greater effects on engine performance.

Fuel characteristics, in particular the tendency for soot formation, also affect internal engine temperatures, with possible implications for performance and durability. Clouds of particulate matter in the high-temperature section of the combustor glow, coloring the flame (technically, they increase its luminosity). Colored flames radiate more heat to the combustor liner, which in extreme cases might overheat. Less soot and a colorless flame promises lower combustor liner temperatures, prolonging maintenance intervals and engine life. Based on research and service experience, engine manufacturers have learned to make combustors that, using standard fuels, now last for thousands of hours before repair or replacement. This suggests that benefits from low-luminosity SPK fuels would probably be small. In the end, conclusions similar to those reached in the 1980s probably still hold: "the impact of fuel type on combustion performance and liner durability is usually small in comparison with the effects of combustor concept, liner geometry, and combustor operating conditions ...."[11]

A4. Fuel Properties and Platform Performance

Fuel specifications normally cover density (lb/gal) and gravimetric energy density (Btu/lb) but not volumetric energy density (Btu/gal); this would be redundant since given any two of these values the third can be calculated. Although gas turbines can be designed to run on almost anything—some have burned powdered coal—kerosene has become the standard because it is dense and inexpensive to make from petroleum fractions not especially suited to other end products.

Designers trade off the weight and volume of fuel carried by aircraft and ground vehicles against other attributes. The range or combat radius of a weapons-carrying "platform," the distance a plane, tank, or helicopter can travel before refueling (or the time it can loiter over a combat zone) depends on the energy content of the fuel. The more space occupied by fuel the less remains for people, electronics, and payload. Energy density, then, is critical, especially for military aircraft. The design compromises are less restrictive for civilian aircraft and not nearly so severe for civilian cars and trucks.

More fuel means more weight, and weight is the enemy of performance, including fuel economy itself.

[11] R. D. Sutton, D. L. Troth, and G. A. Miles, *Analytical Fuel Property Effects—Small Combustors,* NASA CR-174738 (Cleveland: NASA Lewis Research Center, October 1984), p. 123.

These sorts of compromises make more difference for aircraft than for ground vehicles. Since aerodynamic drag rises as the square of speed, a plane at 600 mph consumes a great deal more energy per distance traveled than a car at 60 mph (adjusted for frontal area), and the drag penalty incurred by increasing the size of the plane to accommodate more fuel is far more serious. Weight is also more critical for aircraft. Because a plane that comes off the (virtual) drawing board heavier than intended loses payload capacity, aircraft firms spend thousands of engineering hours shaving weight off this component and that. If a new car comes in at 10 percent over the target weight, it will suffer a fuel economy penalty, may need tires a size larger, could cost a bit more to manufacture—embarrassing but not a debacle. A plane that is too heavy might not get off the runway.

For cars and trucks, refueling is an inconvenience. Interrupting an aircraft flight to refuel is far more time consuming and wastes much fuel since the engines operate inefficiently during approach, landing, taxiing, takeoff, and climb-out. And while air-to-air refueling has long since been routine for military aircraft, it must be carried out away from combat zones and requires a costly fleet of specialized aerial tankers: each of 179 KC-46As the Air Force expects to buy over the next several decades has a projected price tag of nearly $300 million.

Table A2 compares synthetic SPKs to JP-8. The table also lists values for the density and energy content of pure hydrogen and pure carbon to illustrate how properties change at the extremes of carbon-hydrogen ratio. Neither FT nor HEFA synthetics meet the minimum density requirement for JP-8/Jet A/Jet A-1. (Not only do the averages given in Table A2 fall short, so did every sample for which the source documents report individual data.) With some exceptions for cargo and transport planes flying predictable routes, military planes normally take off with full tanks to have the greatest possible margin for contingencies. Lower values of density and of volumetric energy density for the synthetics in Table A2 imply reductions in range or combat radius. These are small and would probably not be viewed as significant enough to bar adoption. Still, the Air Force has explored any number of exotic high density fuels over the years as a means of range extension, and FT and HEFA synthetics move in the wrong direction.

Fuel density and energy content depend on the ratio of carbon to hydrogen, C:H in Table A2, and also on molecular structure. Chemists assign hydrogen, the lightest element, an atomic weight of 1. Carbon has an atomic weight of 12. The more hydrogen relative to carbon in the fuel—the lower the C:H ratio—the lower the density. The synthetics in the table illustrate this relation: they contain less carbon and more hydrogen than JP-8 and exhibit lower measured values of density and volumetric energy density.

The volume occupied by a given molecule depends on how the H and C atoms are arranged, offering another way for chemists and chemical engineers to increase fuel density. DoD has purchased exotic synthetic fuels tailored for high density since the 1960s for specialized applications such as cruise missiles. Smaller is better for cruise missiles, which are powered by jet engines, since more missiles can be fit within the space available in submarines and surface ships (which are volume-limited but not weight-limited) or aircraft (which are always volume-limited and may also be weight-limited). The higher the volumetric energy density of the fuel, the smaller the fuel tank and the smaller the package size for the entire missile; alternatively a bigger warhead could be fitted. Many US cruise missiles burn JP-10, a synthetic with volumetric energy density about 14 percent greater than JP-8 (Table A3). The RJ series of synthetics included in the table, originally intended for ramjets and in a few cases also specified for cruise missiles, offers still higher densities. As the table indicates, RJ-5 is the best of these, not too far from the upper limit for hydrocarbons, which is around 170,000 Btu/gallon.[12] Because RJ-5 freezes at unacceptably high temperatures, it was superseded by RJ-6, a mixture of RJ-5 with JP-10. At least for unclassified applications, RJ-6 too now seems to have become obsolete, probably because of its cost.

The RJ series probed the technical limits of hydrocarbons. Given the effort DoD has devoted to fuels R&D, JP-10 probably represents about as good a compromise as can be expected. The performance attributes of the B-52 bomber, introduced in the 1950s and still the nation's primary intercontinental bomber, illustrate the tradeoffs entailed by density parameters. With its 48,000 gallons of tankage filled with JP-8 and a full payload, 70,000 pounds of ordnance, the B-52H has a nominal combat radius of 4400 miles.[13] Table A4 gives hypothetical figures based on 100 percent FT or HEFA synthetic having properties as in Table A2, with JP-10 included for comparison. Range with the FT and HEFA synthetics decreases by an amount that mission planners would probably consider insignificant. Whether the increased payload—the equivalent of two 5000 pound

[12] W.W. Wimer and E.C. Owens, *Development of Army High-Energy Fuel for Diesel/Turbine Powered Surface Equipment*, Interim Report AFLRL No. 120 (San Antonio: Southwest Research Institute, October 1979).

[13] "US Air Force Fact Sheet: B-52 Stratofortress," <www.af.mil/information/factsheets/factsheet_print.asp?fsID=83&page=1>.

bombs—could be utilized would depend on whether space for munitions or takeoff weight was the binding constraint. With full tanks and JP-10, range goes up but payload drops precipitously. This is not to say, on the other hand, that for other aircraft types for which endurance might be a critical variable—e.g., airborne command—the gains from higher fuel density might not be useful. Cost is the barrier. DLA Energy charges the services $19.98 per gallon for JP-10, more than five time times its price for JP-8.[14] Fuel is an incidental expense for cruise missiles, which cost several million dollars each and fly only once. For most other missions fuel costs may be secondary but will not be immaterial.

As Table A2 showed, finally, pure hydrogen offers much higher gravimetric energy density than hydrocarbons. Yet hydrogen is a liquid only at cryogenic temperatures, close to absolute zero, and even then is low in density and very low in volumetric energy density. For these reasons, liquid hydrogen has been used mostly to power weight-critical rockets (e.g., the space shuttle main engine). At the other extreme in Table A2, solid carbon exhibits much higher volumetric energy content than kerosene. Adding 40 percent carbon particles (by volume) to JP-10 would raise volumetric energy density to about 175,000 Btu/gallon.[15] Metals such as aluminum or boron, and metallic compounds, contain even more energy per unit of volume than carbon. DoD has therefore explored fuel slurries made by mixing powdered solids into kerosene. "In the 1950s, the United States spent hundreds of millions of dollars

[14] "Aerospace Energy Standard Prices for DoD Customers Effective 1 Oct 2012," <www.energy.dla.mil/customer/standard_prices/Pages/AerospaceStandardPrices.aspx>.

[15] Martel, *Military Jet Fuels*, Table 3, p. 11.

to develop boron hydride fuels for jet aircraft. These candidate fuels promised a 40 to 50 percent increase in range as compared to conventional liquid hydrocarbon fuels."[16] Practical difficulties and the associated costs prevented adoption.

[16] Martel, *Military Jet Fuels*, p. 10.

Table A2

Properties of Synthetic Jet Fuels Compared to JP-8

	Density (lb/gal)	C:H Mass Ratio	Final Boiling Temp (F)	Energy Density		Freezing Temp (F)	Flash Point Temp (F)
				Gravimetric (Btu/lb)	Volumetric (Btu/gal) [a]		
Petroleum JP-8 [b]	6.7	7.0	550	18,500	123,900	- 58	125
JP-8 Specification	*6.47-7.0*	*7.5 max*	*572 max*	*18,400 min*	*Not specified*	*-53 max*	*100 min*
Synthetics							
Fischer-Tropsch [c]	6.3	6.5	470	18,970	119,500	- 60	113
Hydroprocessed [d]	6.26	6.6	482	18,990	118,900	- 75	109
Hydrogen (liquid) [e]	0.58	--	*- 423*	*51,600*	*30,400*	--	--
Carbon (solid) [f]	*~ 15.8*	--	--	*14,090*	*223,400*	--	--

Table A2 continued

[a] Unspecified; calculated from density and gravimetric energy density.

[b] Average of 10 samples.

[c] Average of six or more samples (depending on property), except one sample only for freezing and flash point temperatures, from three suppliers (with multiple samples from several).

[d] Average of four samples made from three different feedstocks or feedstock combinations.

[e] Hydrogen is a gas at ordinary temperatures with density and volumetric energy content that depend on confinement pressure. Only at cryogenic temperatures, below - 423 F, does hydrogen become a liquid.

[f] Powdered (carbon black).

Sources: JP-8 - *World Fuel Sampling Program*, CRC Report No. 647 (Alpharetta, GA: Coordinating Research Council, June 2006); Fischer-Tropsch - Clifford A. Moses, *Comparative Evaluation of Semi-Synthetic Jet Fuels* (Alpharetta, GA: Coordinating Research Council, September 2008) and Richard Striebich, et al., *Dependence of Fuel Properties During Blending of Iso-Paraffinic Kerosene and Petroleum-Derived Jet Fuel*, AFRL-RZ-WP-TR-2009-2034 (Wright-Patterson Air Force Base, OH: Air Force Research Laboratory, November 2008); Hydroprocessed - James D. Kinder and Timothy Rahmes, *Evaluation of Bio-Derived Synthetic Parffinic Kerosene (Bio-SPK)* (N.p.: Boeing, November 2009). Hydrogen, carbon - Charles R. Martel, *Military Jet Fuels, 1944 - 1987*, AFWAL-TR-87-2062 (Wright-Patterson Air Force Base, OH: Aero Propulsion Laboratory, November 1987), Table 3, p. 11.

Table A3

High-Density Fuels Compared to JP-8

	Composition	Density (lb/gal)	C:H Mass Ratio	Minimum Energy Density [a]	
				Gravimetric (Btu/lb)	Volumetric (Btu/gal)
JP-8	*Not specified* [b]	*6.5 - 7.0*	*7.0* [c]	*18,400*	*123,900* [c]
JP-10	exo-tetrahydrodicyclopentadiene	7.8 - 7.87	7.5	18,100	141,500
RJ-5	perhydrodi(norbornadiene)	9.0	9.33	17,750	160,000
RJ-6	Mixture of approximately 40 percent RJ-5 with JP-10	8.55 [c]	8.2 [c]	17,850	152,000

[a] Minimum gravimetric energy density specified for JP-8 and JP-10; both gravimetric and volumetric minimums specified for RJ-5 and RJ-6.
[b] Except for impurities.
[c] Typical.

Sources: JP-8, Tables A1 and A2; JP-10 - *Detail Specification: Propellant, High Density Synthetic Hydrocarbon Type, Grade JP-10*, MIL-DTL-87107E (Wright-Patterson Air Force Base, OH: Air Force Petroleum Agency, January 12, 2012). RJ-5, RJ-6 - *Handbook of Aviation Fuel Properties*, CRC Report No. 530 (Atlanta: Coordinating Research Council, 1983).

Table A4

Hypothetical B-52 Payload and Range with Synthetic Jet Fuels

	Fuel Weight (lbs)	Payload (lbs)	Combat Radius (miles)
JP-8	*312,000*	*70,000*	*4400*
FT synthetic	302,000	80,000	4200
HEFA synthetic	300,000	82,000	4200
JP-10	376,000	6000	5000

ABOUT THE AUTHOR

John Alic writes and consults on policy issues related to technology and science. At the congressional Office of Technology Assessment from 1979 to 1995 he directed studies on international competitiveness and technology policy. Alic is author of *Trillions for Military Technology: How the Pentagon Innovates and Why It Costs So Much*, published in 2007, and co-author of *New Rules for a New Economy: Employment and Opportunity in Postindustrial America* and *Beyond Spinoff: Military and Commercial Technologies in a Changing World*. He has also taught at several universities.

Made in the USA
Monee, IL
04 November 2026

67377662R00085